Household Reusable Rainwater Technology for Developing and Under-Developed Countries

Household Reusable Rainwater Technology for Developing and Under-Developed Countries provides insight into household techniques for collecting and treating harvested rainwater safely for both potable and non-potable uses, as well as practices to improve its quality, with numerous real-world case studies and data. It gives a comprehensive, holistic account on the household scale for both developing and under-developed countries.

Improvement mechanisms such as the impacts of first flush, household water treatment techniques, and sedimentation in the harvested water are described in depth together with the advantages and disadvantages of their common practices in developing and under-developed societies. Also discussed is a comprehensive survey illustrating the impact of rainwater sources on the daily life of a carefully selected community from the perspective of its residents.

The book is ideal for students, researchers, academics, water policy providers, and bodies worldwide such as WHO and DFID.

Household Reusable Rainwater Technology for Developing and Under-Developed Countries

Chukwuemeka Kingsley John
and Jaan H. Pu

Routledge
Taylor & Francis Group

LONDON AND NEW YORK

First published 2024
by Routledge
4 Park Square, Milton Park, Abingdon, Oxon OX14 4RN

and by Routledge
605 Third Avenue, New York, NY 10158

Routledge is an imprint of the Taylor & Francis Group, an informa business

© 2024 Chukwuemeka Kingsley John and Jaan H. Pu

British Library Cataloguing-in-Publication Data
A catalogue record for this book is available from the British Library

ISBN: 978-1-032-48890-5 (hbk)
ISBN: 978-1-032-49197-4 (pbk)
ISBN: 978-1-003-39257-6 (ebk)

DOI: 10.1201/9781003392576

Typeset in Sabon
by Deanta Global Publishing Services, Chennai, India

Dedication

Chukwuemeka Kingsley John – My late father, my wife, mother, brothers, nephews, nieces, and friends will always occupy a special place in my heart.

Jaan H. Pu – I would like to acknowledge the contributions and sacrifices of my family, Aigul, Jasmin, and Jeanette, and my parents to my research and during the writing of this book. Their patience and motivation are what help me to get through all the tough and uncertain times.

Contents

Preface

Water is without a doubt a crucial element for life to exist; however, water scarcity (i.e., due to mismanagement and contamination) remains a key challenge for residents in most under-developed and developing countries. While it can take on many different forms, water covers more than 70% of the earth's surface. Although it is believed that the hydrosphere contains more than 1.32 billion cubic kilometres of water, only a tiny percentage of the hydrosphere, roughly 0.3%, is presented in freshwater form, i.e., in rivers, aquifers, springs, and streams, that may be directly useful for daily activities by humans.

Freshwater is a limited resource, and it is at risk of running out while our global population grows. Thus, the methods used to preserve both its quality and quantity become a critical mission to support lives. Harvested rainwater has the potential to provide an extra water resource and hence improve public health. If used and applied correctly, existing technology can support a number of different household infrastructure development models for individual or community-owned local rainwater storage facilities, which arguably can be more effective than centralised treatment and distribution systems.

This book explores harvesting and storage methods for rainwater and the different mechanisms used to improve its quality. Its improvement mechanisms, including the impacts of first flush practices, household water treatment techniques, and sedimentation in the harvested rainwater, have been described in depth together with the pros and cons of their common practices in under-developed and developing societies. Using a designated site at Lagos, Nigeria, the estimation of the health risk of drinking roof-harvested rainwater via the Quantitative Microbial Risk Assessment (QMRA) is discussed in this book. Finally, a comprehensive administered survey on the studied region has been utilised to illustrate the impact of rainwater sources on the daily life of that society, from the perspective of residents. As a key to our motivation, we hope to foresee this book's findings to be formed into a knowledge base, guidelines, and practice method in similar under-developed and developing communities to this studied site.

We encourage enquiries and questions, many of which have helped us to improve.

About the Authors

Chukwuemeka Kingsley John is a passionate learner who currently coordinates the higher education programme in the Department of Construction and Built Environment, which includes Civil Engineering, Building Services Engineering, and Quantity Surveying, at Bath College, UK. Chuks held the same position at University Centre Somerset for about 6 years. His area of speciality is in water, building, and environmental studies, which includes educational research on building, drinking water quality, supply, and hydrodynamics. He is also a certified PRINCE2 manager with experience in managing and supervising several civil and environmental engineering projects. He is an accomplished and highly motivated scientist with extensive research experience in the investigation of water and air quality from different sources in different environments (especially in developing countries). He has extensive experience in contaminated waste management, potential contaminant of concern (PCOC), and hazardous materials investigations involving various industrial processes. Finally, Chuks is committed to working in environmental-related challenges.

Dr Jaan H. Pu is an Associate Professor at University of Bradford, UK. His research focuses on modelling and laboratory approaches to representing various water engineering applications, which include water quality, rainwater, sediment deposition and erosion, naturally compound riverine flow, and vegetated flow. His research outputs have led to several peer-reviewed journal articles (60+), conference proceedings (10+), books (5), and book chapters (2). Jaan has supervised several PhD, industrial, and council-funded projects to investigate river hydrodynamics, sediment transport, and water quality challenging applications. He has been appointed as an associate editor by *Frontiers in Environmental Science* (Impact Factor 5.411) and by *Frontiers in Built Environment* (CiteScore 3.4). He has also served as leading/guest editor of various special issues for MDPI Fluids, MDPI Water, Frontiers in Environmental Science, and Frontiers in Built Environment. He is also a visiting scientist at Tsinghua University, China, and Nanyang Technological University, Singapore.

Abbreviations

ANOVA	Analysis of variance
APHA	American Public Health Association
AWWA	American Water Works Association
AWWS	Average weekly wind speed
CGS	Corrugated galvanised sheet
DAD	Dry antecedent days
E. coli	Escherichia Coli
FF	First flush
FFRW	Free-fall rainwater
HHTTs	Household water treatment techniques
MLR	Multiple linear regression
MPH	miles per hour
MPN	Most probable number
ND	Net deposition
rpm	revolutions per minute
qPCR	quantitative polymerase chain reaction
QMRA	Quantitative Microbial Risk Assessment
RHRW	Roof-harvested rainwater
SSDDS	Sources of drinking water in the dry season
SSDWRS	Sources of drinking water in the rainy season
TD	Total deposition
TRRND	Total roof runoff net deposit
TSS	Total suspended solids
UNICEF	United Nations Children's Fund
WASH	Water, sanitation, and hygiene
WEF	World Economic Forum
WHO	World Health Organization

Chapter 1

Introduction

ABSTRACT

To maintain good health, people need access to clean drinking water. The intake of water from contaminated drinking sources, particularly in under-developed and developing countries, has been linked to a number of diseases. Therefore, one of the first and most important steps to improve a community's quality of life is ensuring that everyone has access to safe drinking water. Researching the effects of harvested rainwater is essential, especially in areas with limited access to pipe-borne clean water. This type of research has the potential to improve public health by improving access to clean water sources. This chapter will focus on the objectives, hypotheses, and structure of the book.

1.1 BACKGROUND

There is no question that the existence of life depends heavily on water. Despite being in various forms, it accounts for more than 70% of the entire surface of the earth, and regardless of habitat, all life forms depend on this life-bearing resource for their continued existence (Daso and Osibanjo, 2012). Although it is estimated that the hydrosphere contains more than 1.32 billion cubic kilometres of water, only a very small portion of the hydrosphere, or around 0.3%, is found to be fresh water in rivers, aquifers, springs, and streams, which is usable for everyday human activities. Seas and oceans make up the remaining 99.7% of the hydrosphere; however, it is not safe to drink water directly from the seas and oceans because they are oversaturated with salt and natural and unnatural contaminants. Desalination is a process of converting salty ocean water into drinkable water, and it is an expensive process to carry out and maintain (Gray, 2010; Chinedu et al., 2011). Millions of people still lack access to safe drinking water, especially in less developed countries. Successful public health interventions are greatly influenced by access to clean water and improved sanitation. Individuals' and communities' health and wellbeing are significantly impacted by the quantity and quality of

DOI: 10.1201/9781003392576-1

the water supply. Access to safe potable water is therefore one of the first crucial steps in improving a community's quality of life (Howard and Bartram, 2003). By managing and supplying sufficient clean and safe water, it is possible to prevent the majority of water-related outbreaks and infections like cholera, dysentery, hepatitis A, legionella, malaria, and typhoid fever (Yang et al., 2012; Tornevi et al., 2013; Herrador et al., 2015; Cinar et al., 2015).

The adverse impact of consuming unsafe water in our daily activities has been demonstrated by several studies. Payment et al. (1999) conducted a prospective epidemiology study to examine the effects of drinking tap water on gastrointestinal health in Quebec, Canada. They concluded that the tap water's noncompliance with current drinking water guidelines was responsible for 14%–40% of gastrointestinal illnesses, with children aged 2–5 being the most affected. Cabral et al. (2009) further revealed that drinking contaminated water is the main factor contributing to the annual death toll of nearly 0.5 million children in India. Research undertaken to evaluate diseases caused by poor sanitation and drinking unsafe water in China found that more than 500 million people lacked access to improved sanitation and more than 300 million people lacked access to safe piped drinking water (Carlton et al., 2012). In that same year, it was claimed that consuming contaminated water and using unimproved sanitation caused 62,800 fatalities and 2.81 million disability-adjusted life years (DALYs) nationwide, with children under the age of 5 bearing the brunt of 83% of the risk burden (Carlton et al., 2012). Poor sanitation, hygiene, and water quality contributed to nearly 1.6 million fatalities worldwide in 2006 (3.7% of all DALYs and 3.1% of all deaths), primarily from infectious diarrhoea (Ashbolt and Kirk, 2006). These literatures reiterated the consequences involved in ingesting unclean and unsafe water in different parts of the world.

Access to safe drinking water is fundamental to human health. Prior to the Millennium Development Goal (MDG) 2015 deadline, the World Health Organization (WHO) declared in March 2012 that the world had achieved the MDG of halving the percentage of people without sustainable access to clean drinking water (WHO/UNICEF JMP, 2012). According to statistics, 95 nations have achieved the sanitation target, 147 countries have reached the drinking water target, and 77 countries have achieved the MDG targets for both drinking water and sanitation (Bain et al., 2020). This is an improvement, because only 56 countries had met the MDG target for both improved drinking water and sanitation in 2012. Furthermore, 2.6 billion people had access to better drinking water sources between 1990 and 2015, while 2.1 billion individuals have access to better sanitation globally. Despite this improvement, 946 million people still use open defecation, and 2.4 billion continue to use unimproved sanitation facilities (WHO/UNICEF JMP, 2015). Also, it was estimated that about 663 million people globally still ingest unsafe water from unimproved drinking water sources, including surface water, springs, and unprotected wells. In addition, most of the

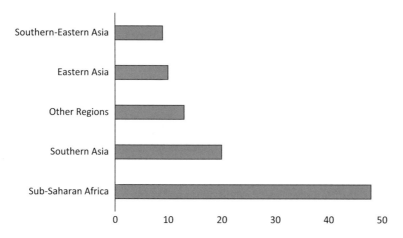

Figure 1.1 World population by region in 2015, without access to better water sources (WHO/UNICEF JMP, 2015)

population who do not have access to improved drinking water currently reside in just two of the continents of the world, mostly in less economically developed areas (Figure 1.1). At this rate, it will take four times as much effort to achieve drinking water, sanitation, and hygiene goals by 2030 (Bain et al., 2020). Nevertheless, due to inadequate infrastructure and surveillance, many nations, particularly developing ones, are unable to uphold these water supply criteria. According to the most recent estimates from the WHO/UNICEF Joint Monitoring Programme (JMP) for Water Supply, almost 2 billion people do not have access to safely managed drinking water (SMDW) services. The majority live in sub-Saharan Africa (with 747 million people) and Central and South Asia (with 768 million people) (Bain et al., 2020). The "Rapid Assessment of Drinking Water Quality" project by WHO and UNICEF in support of the MDG gave encouraging results in the improvement of drinking water sources, but some studies like Bain et al. (2014) reported that over a quarter of samples from improved sources were contaminated with faeces in 38% of 191 studies from sub-Saharan Africa, Southern Asia, Latin America, and the Caribbean. A similar study by Onda et al. (2012) proposed that the issue of contamination of improved water sources has to be understood and managed with more care. Water sources were more likely to be contaminated in rural and low-income communities. They stressed the significance of modifying the monitoring system and creating new MDGs. Although strict laws have been passed and regulatory organisations have been established in some nations to make sure that the standards are being upheld and enforced, the administration and implementation of the law present enormous difficulties (Weinmeyer et al., 2017; Bain et al., 2020).

In the 2012 MDGs report, Nigeria was listed as one of the world's top ten most populated nations lacking access to an improved drinking water source. It was further reported in 2012 that Nigeria will meet neither the MDG target for the projected use of improved sanitation facilities, nor the target for improved drinking water sources by 2015 (WHO/UNICEF JMP, 2012). The 2015 MDG report, however, stated that Nigeria had met the target for improved drinking water but that limited or no progress has been made for improved sanitation facilities. Despite Nigeria narrowly meeting the target for improved drinking water, the report showed that more of these improvements were recorded in the urban areas, therefore making it imperative to investigate the impact of harvested rainwater, especially in rural areas where there is limited access to community pipe-borne treated water. Harvested rainwater has the potential to provide an extra water resource and hence improve public health. If applied correctly, the existing technology can also support a number of different infrastructure development models that may be more effective than centralised treatment and distribution systems. Such models include individual or community-owned local storage facilities.

WHO lists rainwater as an improved source of drinking water. Improved water is defined as "one that, by nature of its creation or via active intervention, is secured from external contamination, most especially from contamination with faecal matter" (WHO/UNICEF JMP, 2014). Clean rainwater has been considered to be a safe and appropriate source of drinkable water (Asthana and Asthana, 2003; Nsi, 2007; Olowoyo, 2011), and it is commonly used in the underdeveloped and some developing countries of the world. Treated pipe-borne water is lacking in most communities in the less economically developed countries, and therefore free-fall harvested rainwater and roof-intercepted or harvested rainwater can be a main supply of drinking water, especially during the rainy seasons (Olabanji and Adeniyi, 2005). However, the results from some studies showed that the quality of natural roof-harvested rainwater is poor in general (Ward et al., 2010; Islam et al., 2011; Gikas and Tsihrintzis, 2012).

One of the apparent limitations in the appropriateness of rainwater as a source of drinking water is the potential sources of microbial contamination, particularly from the surrounding environment. The hypothesis that the quality of harvested rainwater is hugely dependent on local environmental factors, not just sanitation systems, will be investigated and justified in this book. Furthermore, the rate of roof-top deposition of particulate matter and its contamination with pathogens has been found to be influenced by the extent of unpaved roads and non-vegetated areas. The increased volume of wind deposit can also provide a transport route for bacteria associated with the deposits (John et al., 2021a). Thus, it is imperative to investigate the fate of the microbes in storage tanks to identify specific mechanisms for water quality improvement, such as reducing sedimentation impact. In this book, separate field survey and experimental phases are utilised to provide the input data for a Quantified Microbial Risk Assessment model for stored rainwater analysis. This model can also be used to investigate the impact of different infrastructure development scenarios on public health.

Rainwater quality analyses have been done by different researchers in some parts of the world, some of which were performed in Nigeria. Despite some research presenting the inconsistent nature of microbial quality in rainwater, other studies show that harvested rainwater is a significant water resource in developing countries. Results from studies piloted by Ariyananda and Mawatha (1999), Pushpangadan et al. (2001), and Handia (2005) in Sri Lanka, India, and Zambia, respectively, showed that the microbial quality of other sources of drinking water like surface water and shallow groundwater were poorer than that of stored rainwater. To appropriately investigate the methodology to improve the quality of harvested rainwater in different situations, this book will first identify the research gaps and invest effort to understand them. The following knowledge gaps will be investigated in this book:

1. There has been very limited research effort on atmospheric deposition in Nigeria. Therefore, this book will illustrate the impact of rate of deposition of particulate matter on rainwater and its quality. One study area, the Ikorodu Governmental Area at Lagos, will be explored in detail.
2. This book will continue to study the seasonal variation of the deposition rate of particulate matter and the quantity of bacteria (i.e., Total Coliform and *E. coli*) in the sediments of the matter in both rainy and dry seasons.
3. Next, it will investigate the influence of wind on the deposition of particulate matter.
4. The World Health Organization (WHO) listed rainwater as an improved source of drinking water. Hence, rainwater sanitation and the impact of the surrounding pavement type (in terms of paved and unpaved areas) on the quality of rainwater will be investigated.
5. Currently, most studies in Nigeria focus on the general analysis of rainwater without controlling any major factors (such as the influence of tank positioning, tank size, and season). This book will analyse two aspects of rainwater quality:
   • The level of contamination due to environmental factors.
   • The subsequent improvement due to storage.

1.2 AIM, OBJECTIVES, AND HYPOTHESIS

This section outlines the description of the overall objectives and the hypotheses of the study. The specific and general definitions used in this book are presented in Appendix A. The aim of this book is to investigate and develop a model that can explore the impact of different infrastructure development scenarios on public health. This was achieved by determining the health risk involved in using roof-harvested rainwater as a potable source of water, where a targeted study area has been investigated (Ikorodu Governmental Area at Lagos). The Quantitative Microbial Risk Assessment process was used to evaluate the infection probability by pathogenic bacteria.

The aim of this book was achieved via the following objectives:

- Determine the rate of particulate matter deposition in a single property and the impact of variable factors on deposition in both the dry and rainy seasons using the study area and identify the range of the bacteria associated with particles in the deposited solids from the roof.
- Compare the rate of deposition of particulate matters and their associated bacteria in different locations in the study area.
- Investigate the rate of improvement caused by the sedimentation process for both the physical and microbial quality of stored rainwater, and study the effect of household treatment techniques such as chlorine, boiling, and alum on the particle settling and microbial load.
- Determine the water use strategies (i.e., the sources of drinking water and domestic usage of different water sources in both seasons) and the water and sanitation infrastructure in the study area via a well-structured questionnaire which was evenly distributed in the study area.
- Investigate the impact of first flush in improving the quality of roof-harvested rainwater from five differently constructed roof types (including asbestos) in a household.
- Estimate the health risk from drinking roof-harvested rainwater via the Quantitative Microbial Risk Assessment process by the administered survey data and from the enumerated *E. coli* values from the rainwater storage tank.

The key hypothesis for the proposed model within this book is that people in areas with poor sanitation are at a high risk of consuming contaminated water containing pathogenic bacteria. This is due to increased transport rates and levels of bacteria in the environment. However, the amounts of bacteria that are ingested can be reduced by designing collection and storage systems that speed up the sedimentation process. The majority of bacteria identified are associated with particles with a diameter in the range of 10 μm to 100 μm (John et al., 2021b). The percentage and categories of the people who collect and drink roof-harvested rainwater in the study area will be reported. The four categories of people who collect and consume roof-harvested rainwater were examined in this book, and these four categories (which include individuals who utilised various types of household treatment techniques) were acquired during the questionnaire stage of the first phase of field work.

1.3 BOOK STRUCTURE

This book is multidisciplinary in nature, as rainwater is impacted by various geographical, physical, biological, and chemical processes. Hence this book will be covering the following:

- The magnitude and particulate deposition rate are specific to a given area. Therefore, the rates of roof-top deposition and free-fall particulate matter over time, considering the influence of environment (paved roads, drainage infrastructure, and method of sanitation); distance from the source; rainfall pattern; seasonal impact; and wind speed will be described. It is crucial to understand the rate of any accumulation as the resulting quality of rainfall run-off will vary significantly with the length of the antecedent dry period (Chapters 4 and 5).
- The impact from different size fractions of the deposit will be investigated to find their different levels of contamination with pathogens. This partitioning will determine the proportions of the bacteria that are associated with mineral or organic particulate matter and the proportion of bacteria that can be removed via simple sedimentation or household treatment techniques (Chapters 4, 5, and 6).
- The impact of first flush in improving the quality of roof-harvested rainwater will also be illustrated (Chapter 6).
- The estimation method of the health risk of drinking roof-harvested rainwater via the Quantitative Microbial Risk Assessment will be described in detail (Chapters 5, 6, and 7).

The structure of the book is depicted in Figure 1.2 along with the interconnective nature of the different chapters. Chapter 2 sets out the context in which the study was carried out. The relevant literature are reviewed on the atmospheric deposition of particulate matter and the possible risk associated with the water quality in various rainwater harvesting systems. This chapter also presents sedimentation as a mechanism for water quality improvement and further outlines the various stages in Quantitative Microbial Risk Assessment. A synopsis of the existing literature relating to the rainwater harvesting systems around the world with an emphasis on less economically developed countries and a brief description of the study area are provided.

Chapter 3 details the available experimental methods, i.e., for measuring net roof-top deposition, total atmospheric deposition, and other relevant parameters. These methods are referenced throughout the experiments in the different chapters. Chapter 4 covers the rate of deposition of particulate matter from a single building's roof-top. In this chapter, the measured total atmospheric deposition, net deposition, dry antecedent days, wind speed, and rain intensity are explored and discussed in both rainy and dry seasons. Also, the utilised serial filtration of the deposit to determine the size ranges

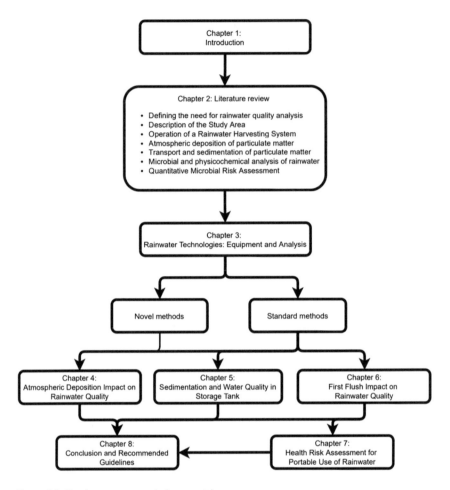

Figure 1.2 Book structure and chapter titles

of the deposited solids is described. Chapter 4 also studies the rate of total atmospheric deposition from four different locations with different features.

Chapter 5 inspects the quality of the harvested rainwater from different levels within different periods in storage tanks. This has been done for different rain events in a domestic household in the study area. The influence of the different household treatment techniques applied by residents to harvest rainwater and the quality of free-fall rainwater compared to roof-harvested rainwater are discussed. The fractionation of the solids and the impact of sedimentation in the storage tanks are also evaluated in this chapter. Chapter 6 investigates the impact of first flush on roof-harvested rainwater in the study area. The first flush practices are discussed to outline their effectiveness. Chapter 7 gives a detailed description of the rainwater quality questionnaire practice, whereby an administered survey was conducted with the residents of the study area. The obtained data from the questionnaires have been described and used to develop the Quantitative Microbial Risk Assessment.

The final chapter, Chapter 8, summarises the study outcomes of this book and other literatures and recommends ways that this research can be progressed in the future. The list of references cited is presented at the end of the book. A list of design calculations that are related to the experimental apparatus and other results which have been presented in the body of the book are set out in Appendices B and D. Appendix E presents the administered rainwater questionnaire form and details. Lastly, Appendix F presents the Most Probable Number (MPN) table used for Colilert results in water quality discussion, while Appendix G displays rain event characteristics during the sampling trials.

1.4 SUMMARY

This introductory chapter presents a rationalisation to the background of this book. The general objectives of the research, the structure of the book, and the knowledge gap as well as the contribution to knowledge field have been defined. An executive summary of the contents in the forthcoming chapters is given. The hypotheses used in this book have also been laid out and discussed in this chapter.

Chapter 2

Literature Review on Rainwater Quality

ABSTRACT

This chapter presents a summary of previous and current research on harvested rainwater quality and systems around the globe, with a focus on developing and under-developed countries. The methodological approaches and theoretical contributions in this field are discussed. A comprehensive description of the chosen study area and the justification for the need to assess and monitor the quality of harvested rainwater in this area is provided. Rainwater harvesting technologies in developing countries are further explored and described, and the benefits and difficulties of the most used techniques are highlighted. Local environmental conditions are known to have an impact on rainwater quality due to varying rates of particulate matter deposition from the atmosphere. This concept of particulate matter deposition is also covered within this book.

2.1 DEFINING THE NEED FOR ANALYSING THE QUALITY OF DRINKING WATER

Fresh water availability is threatened by a number of issues, including pollution and climate change. It is crucial to fund research and technological advancements that can lessen the impact of climate change and protect the supply of fresh water. A simple sustainable method of collecting and fully utilising fresh water is through the harvesting of rainwater. Utilising harvested rainwater has many advantages, which include (i) reducing roof runoff, which may frequently deteriorate the health of creek ecosystems; (ii) providing an alternative source of water supply during periods of water scarcity; and (iii) lowering the demand on the main water supply sources. Irrespective of these advantages, rainwater has not been extensively used for ingestion due to a dearth of knowledge on the prevalence and risk from microbiological, chemical, and physical pollutants (Ahmed et al., 2011). The need to analyse the quality of potable water is continuously reiterated in the literature (WHO, 2011, Ahmed et al., 2012; Dean and Hunter, 2012). One of the goals of this book is to quantify the risk factors and offer recommendations for mitigating

DOI: 10.1201/9781003392576-2

risks associated with the adoption of rainwater harvesting technologies in economically less developed countries.

In many nations around the world, the practice of collecting rainwater for home consumption is getting more attention and hence is well documented by numerous scholars. Consequently, its associated health risk and quality have been evaluated by different studies (Gould, 1999; Lye, 2002, 2009; Ahmed et al., 2009; Dean and Hunter, 2012; John et al., 2021c). *E. coli* enumeration is typically used to assess the bacteriological quality of tank-stored rainwater samples since they are typically found in the stomach of warm-blooded animals (Pinfold et al., 1993; Spinks et al., 2006; Sazakil et al., 2007; Ahmed et al., 2012). Different authorities (such as WHO, UNICEF, USEPA, the Nigerian and other national authorities) have established the guideline limit to assess the suitability of water to be used for potable purposes. Most of these authorities specified that the amount of *E. coli* should be 0 in 100 mL of water (WHO, 2004; SON, 2007; Ahmed et al., 2011, 2012).

Drinking contaminated water is linked to several diseases, including schistosomiasis, trachoma, lymphatic filariasis, intestinal nematode infection, and diarrheal illness (Pruss et al., 2002; Gray, 2010; WHO, 2011). Diarrhoea, caused by microbial contamination, is one of the most common disease burdens associated with drinking contaminated water (WHO, 2011). Diarrhoeal disease is the second prominent cause of death in children under 5 years and it kills approximately 1.6 million children each year. Additionally, diarrhoea kills more young children compared to deaths from measles, malaria, and acquired immunodeficiency syndrome (AIDS) combined (Table 2.1) (Manetu et al., 2021). Approximately 2.5 billion cases of diarrhoeal disease occur every year globally, with the major impact being in economically less developed countries. Lozano et al. (2012) study on regional and global mortality from 235 causes of death for 20 age groups between 1990 and 2000 showed that diarrhoeal disease is the second biggest cause of years of life lost in sub-Saharan Africa (behind malaria), and this was attributed to the lack of access to improved safe drinking water sources and poor sanitation facilities. This book will contribute to ameliorating the dearth of knowledge on the quality of harvested rainwater using a case study at Ikorodu, Nigeria.

India has the highest prevalence of paediatric diarrhoea (refer to Table 2.2), with more than 380,000 children dying from its symptoms (Manetu et al., 2021). However, India has made significant progress in lowering the death rate for children under the age of 5. This accomplishment can be attributed

Table 2.1 Global cause-specific mortality rates for children under 5 (Manetu et al., 2021)

Disease	Pneumonia	Diarrhoea	Neonatal causes	Malaria	Measles	AIDS	Others
% of deaths	17	16	37	4	4	2	4

Table 2.2 Cases of diarrhoea among children under 5 years old by Country Manetu et al. (2021)

Rank	Country	Country-level total of annual child fatalities from diarrhoea
1	India	386,600
2	Nigeria	151,700
3	Democratic Republic of the Congo	89,900
4	Afghanistan	82,100
5	Ethiopia	73,700
6	Pakistan	53,300
7	Bangladesh	50,800
8	China	40,000
9	Uganda	29,300
10	Kenya	27,400
11	Niger	26,400
12	Burkina Faso	24,300
13	United Republic of Tanzania	23,900
14	Mali	20,900
15	Angola	19,700

to the creation and expansion of numerous universal programmes, such as vaccine initiatives. Despite the reduction, India still has a significant share of deaths attributable to diarrhoea (Lakshminarayanan and Jayalakshmy, 2015; Manetu et al., 2021). The prevalence of diarrhoea is still very high throughout Africa. An estimated 30 million instances of severe diarrhoea and 330,000 fatalities were attributed to diarrhoeal illness in Africa in 2015 alone (Reiner et al., 2018; Manetu et al., 2021). In Africa, a lack of access to sufficient clean water, poor sanitation, and poor hygiene habits worsen diarrhoea incidence and severity. Furthermore, diarrhoeal diseases continue to be one of Kenya's key public health issues, with the mortality rate of children under 5 years due to diarrhoea being extremely high (approximately 16% higher than deaths from HIV and malaria combined). Furthermore, every child in Kenya under the age of 5 experiences an average of three episodes of diarrhoea per year (KDHS, 2014). Over 50% of hospital visits in Kenya are for illnesses that are related to water, sanitation, and hygiene practices, which are mostly attributed to poor hygiene standards, inadequate sanitation infrastructure, and a lack of access to safe and clean drinking water (Manetu et al., 2021).

According to the Institute for Health Metrics Evaluation, an estimated 6.8% of disability-adjusted life years (DALYs) in 1990 and 3.7% in 2000 were attributed to the use of unimproved drinking water sources and poor sanitation facilities, whereas an estimated 0.9% of DALYs were observed in 2010 (Lim et al., 2012; Clasen et al., 2014). After the researchers compared the mortality rate from diarrhoeal diseases between 1990 and 2010, their results showed that a 41.9% reduction in mortality was observed where the affected population reduced from 2.5 million to 1.4 million (Lozano et al., 2012; Clasen et al., 2014). The decline over time in the death rate from diarrhoeal illnesses and DALYs was credited to the improvement in the worldwide Millennium Development Goals (MDG) target for improved sanitation and drinking water. Therefore, continuous effort is required to enhance access to improved sources of drinking water. Since harvested rainwater offers huge potential in this area, it is important to analyse and improve its quality.

Analysing the quality of drinking water is important for detecting the presence or absence of infectious micro-organisms, including bacteria and protozoa. Giardiasis is a contagion that affects the small intestine, and it is caused by a microscopic protozoan called Giardia lambia (Coskun et al., 2010). Giardiasis is one of the infections caused by ingesting water and food contaminated by protozoal bacteria and by contact with infected people. In economically less developed nations, where there is overpopulation, poor access to safe water, and inadequate sanitary facilities, this disease is common (Caccio, 2004; Gray, 2010; Coskun et al., 2010). Painter et al. (2015) investigated the variation in the reported cases of Giardiasis disease in the United States of America between 2005 and 2012 and observed a slight reduction in case rates in 2011 and 2012 from the relatively stable rates during 2005–2010. The range of the cases for 2005–2010, 2011 and 2012 per 100,000 population are 7.1–7.9, 6.4, and 5.8, respectively. The decline in case rates was attributed to altered waterborne transmission pathways and increased acceptance of prevention measures. Continuous educational initiatives to lower exposure to harmful ingesting water, as well as prevent person–to-person transmission, is advised to reduce the spread of Giardiasis (Painter et al., 2015). Findings from a cross-sectional study carried out in the Lipis region of Pahang state, Malaysia, to determine the prevalence and current risk factors related to various intestinal parasite species among school children revealed that 71.4% of the schoolchildren had poly-parasitism, while 98.4% of the 498 students were found to be infected with at least one parasite species (Al-Delaimy et al., 2014). The examined parasites in that study are *Cryptosporidium spp*, *Entamoeba spp.*, *Giardia duodenalis*, hookworm, *Ascaris lumbricoides*, and *Trichuris trichiura*, and their respective prevalence of the infections are 5.2%, 14.1%, 28.3%, 28.3%, 47.8%, and 95.6%. Further examination of the data revealed that the main risk factors for intestinal poly-parasitism in these children were poor personal cleanliness, a lack of access to improved drinking water, and inadequate sanitation facilities. It was concluded that intestinal poly-parasitism is highly prevalent among children in the area of Malaysia under investigation. To reduce the occurrence and concerns of these infections, it was advised that effective and long-lasting control measures must be implemented. These measures include offering adequate health education that focuses on good personal hygiene practices and proper sanitation, as well as supplying safe drinking water (Al-Delaimy et al., 2014). Obtaining water of good quality and implementing the necessary treatments or prevention measures will go a long way towards eradicating or significantly reduce waterborne and water-related diseases, especially in young children, who have a weaker immune system (Al-Delaimy et al., 2014; Painter et al., 2015). Therefore, it is also crucial to frequently check the quality of drinking water.

Water contamination from microbes and other sources is a problem that affects people all around the world. The extent of challenges posed by this issue varies between nations. While some developed countries in the world tackle the issue of the sustainability of water and degraded water quality, other developing countries find it difficult to get access to improved drinking water sources and sanitation facilities (Ariyananda and Mawatha, 1999; DFID, 2015). The 2015 MDG reports on improving drinking water sources further underlined this need. The report stated that the global use of improved drinking water sources rose from 76% to 91% between 1990 and 2015, thus estimating that approximately 6.6 billion people currently use improved drinking water sources. While over 90% of developed regions have access to improved drinking water sources, only 69% of the least developed countries have access to it, with the high percentage of residents in Oceania and sub-Saharan Africa continuously drinking water from surface water sources, which are regarded as being an unimproved source. These surface water sources include ponds, rivers, irrigation canals, and lakes (WHO/UNICEF JMP, 2015).

The same water quality issues affecting developing and developed countries are also applicable to Nigeria. Although the MDG progress assessment reported that Nigeria met the 2015 goal of improved access to drinking water, statistics from the report further showed that this goal was mostly achieved in urban areas as compared to rural areas. The data in Tables 2.3 and 2.4 show the disparity of the accomplishment levels of the goals between rural and urban settings. The report further showed that Nigeria had not made significant progress in improving sanitation facilities (WHO/UNICEF JMP, 2015). Despite the accomplishments of the MDG in terms of drinking water in Nigeria, it is important to continuously look for ways to provide more improved drinking water sources and access to them. In addition, it

Table 2.3 2015 Progress report towards the MDG drinking water target for Nigeria (WHO/UNICEF JMP, 2015)

Year	% of total improved DW in urban areas	% of total unimproved DW in urban areas	% of total improved DW in rural area	% of total unimproved DW in rural areas	Progress towards the MDG target	% of the 2015 population that gained access since 1990
1990	76	14	25	75	Met target	48
2015	81	19	57	43		

*Note: % and DW implies percentage and drinking water, respectively

Table 2.4   2015 Progress report towards the MDG sanitation facilities target for Nigeria (WHO/UNICEF JMP, 2015)

Year	% of total improved SF in urban areas	% of total unimproved SF in urban areas	% of total improved SF in rural area	% of total unimproved SF in rural areas	Progress towards the MDG target	% of the 2015 population that gained access since 1990
1990	38	62	38	62	Limited or no progress	9
2015	33	67	29	71		

*Note: % and SF implies percentage and sanitation facilities, respectively

was found that over 25% of the Nigerian population did not have access to an improved drinking water source despite the achievement of the country's MDG (WHO/UNICEF JMP, 2015). This book will therefore investigate the recent quality of rainwater as an improved source of drinking water in a rural area of Nigeria.

2.2 DESCRIPTION OF THE STUDY AREA

Nigeria is situated in West Africa between latitude 4° and 14° North of the Equator, and between longitude 2° 2' and 14° 30' East of the Greenwich meridian. Nigeria is surrounded to the north by the Republics of Niger and Chad, to the south by the Atlantic Ocean, to the west by the Republic of Benin, and to the east by the Republic of Cameroon. It has a significant land area of about 923,770 square kilometres (Helmer and Hesanhol, 1997; BGS, 2003). The country of Nigeria comprises 36 states and its Federal Capital Territory, Abuja (Figure 2.1), and these are made up of 774 local government areas. Nigeria is the most populated nation in Africa, with about a quarter of the total population situated in sub-Saharan Africa. With a population of over 200 million people, Nigeria has the sixth largest national population in the world (United Nations, 2022). Despite being a lower middle-income country with the world's 27th largest economy (United Nations, 2022), Nigeria is positioned as one of the top five African countries with the highest gross domestic product or GDP (United Nations, 2022). However, Nigeria has a significant burden of disease linked to environmental, infrastructural, and sanitary problems in terms of health and social infrastructure (Al-Amin, 2013; Juniad and Aigna, 2014).

In Nigeria, there are three distinct climate zones: equatorial in the south, tropical in the centre, and arid in the north. The amount and quality of Nigeria's water resources are impacted by the country's climate. This occurs due to two main wind systems: the hot, dry, and dusty wind that drifts from the north-east over the Sahara Desert and brings dry weather and dust-laden

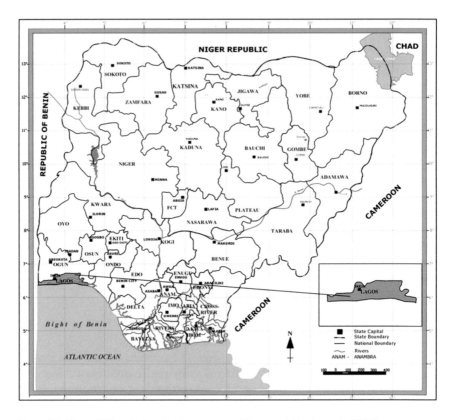

Figure 2.1 Map of Nigeria showing Lagos state (Source: Ajibade et al., 2014)

air; and the relatively cold, moist, monsoon breeze that drifts from the south-west over the Atlantic Ocean, which brings rains. Nigeria experiences an average annual rainfall of 1180mm, with regional variations ranging from 250mm in the north to over 4000mm in the south. In Nigeria, there are two distinct seasons: the rainy season, which lasts from April to September/October; and the dry season, which lasts from November to March. The average yearly temperature is between 26°C and 31°C (78 and 88 °F) (Helmer and Hesanhol, 1997; Carter and Alhassan, 1998; Alagbe, 2002; BGS, 2003). According to Fagbenle and Karayiannis (1994), the average wind speed in Nigeria is between 2 and 4 m/s, with the southern and northern regions experiencing peak mean winds of about 3.5 to 7.5 m/s. Consistently, Adekoya and Adewale (1992) found that the average wind speed in Nigeria varied from 1.5 to 4.1 m/s. The yearly average wind speed in the northern regions of the nation is equal to or greater than 2.5 m/s, whereas mean wind speeds of less than 2.5 m/s have been recorded in the southern regions. It is understood that the wind speed fluctuates with time and is not a continuous factor. In both the rainy and dry seasons, it was found that the rate of deposition and subsequent re-suspension of particulate matter on rooftops is influenced by the local wind speed (John et al., 2021b).

Surface water (such as streams, lakes, rivers, and estuaries), pipe-borne water, boreholes, dams, hand-dug wells, and rainwater collection are the primary sources of drinking water in Nigeria (FGN, 2000; Longe et al., 2010). The country's six geopolitical zones were estimated to have an annual water supply of 224 billion cubic metres, but the aquifer's groundwater resources have not been quantified in the literature (Handidu, 1990; Longe et al., 2010). Water is mostly used for domestic activities, farming, raising farm animals, irrigation, and fishing activities in a typical traditional Nigerian setting. The primary domestic uses of water include bathing, washing, and drinking (Kuruk, 2005; Longe et al., 2010). It was estimated that about 96 million Nigerians (about 48% of population) use surface water for their domestic needs (Longe et al., 2010), while 20% of the population (about 40 million) depend on harvested rainwater for their needs (FGN, 2000; Longe et al., 2010). This illustrates the importance of improving the quality of harvested rainwater, since millions of people depend on it for their daily needs (including potable usages). This book will further investigate the quality of free-fall rainwater and roof-harvested rainwater.

The Nigerian government is divided into three tiers: Local, State, and Federal Governments. The responsibility for providing clean water and managing water resources is shared by Nigeria's three tiers of government, which has historically led to misunderstandings and inefficiency (Longe et al., 2010; USAID, 2012). While the Federal Ministry of Water Resources (FMWR) is required to develop policies, collect information, and coordinate and monitor the development of water supplies as well as provide financial support for research and development, the state water agencies are tasked to develop, operate, maintain, control, and improve the quality of the urban and semi-urban water supply (Ikelionwu, 2006; WSMP, 2008; Longe et al., 2010; Arowolo et al., 2012; USAID, 2012). Only a few of the 774 Local Government Authorities have the expertise and human resources to address the sector, despite their responsibility to build, maintain, and operate rural water supply schemes and sanitation facilities in their respective regions (Longe et al., 2010; USAID, 2012). Nigeria has 37 water agencies, with the key one in the capital and one in every state. The water agencies were established as corporate entities that belonged entirely to the state government, but they are regularly run in accordance with civil service regulations. The introduction of service public–private partnership has resulted in restructuring in four of the 37 water agencies, notably the state water agencies in Ogun, Lagos, Kaduna, and Cross River (USAID, 2012).

Despite the responsibility of these organisations to offer clean water, the majority of their websites and databases have few or no records of water quality. The National Water Research Institute (NWRI), one of the government's parastatals that is responsible for the effective water quality assessment and monitoring, reported that "after two decades of inactivity, the Institute has initiated a project aimed at re-establishment and modernizing the National Water Resources Data Management System" (NWRI, 2012). The need for organised and standardised country-level data monitoring makes it easy to either assess the current state of water quality or to work with these records to estimate and predict future issues relating to water quality in the country. This book will further explore and contribute to the currently limited data on rainwater quality in Nigeria.

The study area is the Ikorodu Local Government Area of Lagos. Ikorodu is a city which shares a boundary with Ogun state and is situated in northern region of Lagos, approximately located at latitude 6° 36' north of the equator and longitude 2° 30' east of the Greenwich meridian (Maplandia, 2015). This area was chosen because it is one of the fastest developing areas in Lagos and is known for high adoption of rainwater harvesting for potable use, especially during the rainy season (Longe et al., 2010; Ukabiala et al., 2010). The central sections of Ikorodu are home to a variety of land use activities including residential buildings, business and retail establishments, governmental and private organisations, and other land uses, e.g., farming. It was found that more than 43.5% of all buildings were used for commercial purposes in the study by Bello (2007a). These land uses result in daily traffic, business, and other active anthropogenic activities. A broad assessment of the area's overall developments showed that the area is not well-planned and might not be able to meet the expanding needs brought about by new development activities. Finally, except for the main route connecting the city centre to the central regions of Lagos state, the majority of the roads are not paved nor maintained well (Bello, 2007b).

2.3 OPERATION OF RAINWATER HARVESTING AND STORAGE SYSTEMS

Most rainwater harvesting systems are designed to collect rainwater from roofs and store it in tanks or other containers. These rainwater systems vary in design from a straightforward barrel at the end of a downspout to a composite potable or multiple end-use system that uses a sizeable water storage tank. Jones and Hunt (2008) outlined four essential elements that make up a typical rainwater collection system. These elements can be described thus:

- The gutter system or the network of gutters makes up this system, where the rooftop runoff from these systems is collected and directed into the storage tank or cistern.
- The overflow pipe collects extra runoff and channels it out of the storage tank in a controlled way.

- The outflow pipe gathers water from the storage tank's bottom for usage, and it is occasionally coupled with a pump.
- The cistern/water storage tank serves as a reservoir to hold runoff from the rooftop for later use. This component is very important to the rainwater harvesting system, and it is usually selected and located based on its anticipated needs. Depending on the user and location, the water storage tank can be kept above the ground or underground. A key component of the rainwater system that resembles the sedimentation basin in a full-scale water treatment facility is the water storage tank. In each instance of sedimentation basin in the full-scale water treatment plant or storage tank of the rainwater harvesting system, there is a need for aeration, disinfection, and sedimentation to take place. The means by which particles settle out of suspension are vital to the design and operation of tanks.

In the full-scale water treatment process, most of the remaining suspended solids after the fine screening stage are usually colloidal in nature. These solids may include clays, metal oxides, proteins, and micro-organisms, and they are negatively charged, repel each other, and prevent aggregation and settlement. These colloidal repelling particles are either hydrophobic (i.e., do not absorb water) or hydrophilic (i.e., do absorb water), thus making them unstable and remain in suspension. A coagulant can be used to induce particle clusters due to their positive charge, thereby stabilising the particles so that they agglomerate and settle (Gray, 2010). The most common coagulants include alum ($Al_2(SO_4)_3$ $14H_2O$), aluminium hydroxide, ferric chloride, and ferric sulphate. Alum reacts with the alkalinity in the water to produce insoluble aluminium hydroxide floc ($Al(OH)_3$), according to Equation 2.1:

$$Al_2(SO_4)_3 .14H_2O + 3\ Ca(HCO_3)_2 \rightarrow 2Al(OH)_3 + 3CaSO_4 + 14H_2O + 6CO_2 \tag{2.1}$$

If alkalinity is insufficient in the water, lime in the form of calcium hydroxide can be added to obtain Equation 2.2

$$Al_2(SO_4)_3 .14H_2O + 3\ Ca(OH)_2 \rightarrow 2Al(OH)_3 + 3CaSO_4 + 14H_2O \tag{2.2}$$

The elements described above depict a typical basic rainwater harvesting system in a developed country, but there are several more common ways of harvesting rainwater in other parts of the world, especially in developing and under-developed countries. One of these includes putting the collection vessels under the roof during rainfall, and then transferring the harvested rainwater to the storage vessel, which may be kept either inside or outside the house or both. The collection vessel is also known as the temporary storage vessel. The purpose of the collection vessel is to receive water from the roof and to temporarily store the water while the rainfall event is occurring. The potential impact of the collection vessel on the quality of the harvested water is dependent on the hygiene and sanitation practice of the household in terms of the collection vessel.

As the name implies, the storage vessel is used to store water for future use. This vessel receives the water that was initially gathered by the collection vessel for storage until the water is used. The potential impact of such storage on water quality is also dependent on the household hygiene practices towards the storage vessel, and the conditions in terms of the surroundings of the storage vessel. Hygiene and sanitation practices include how close the vessels are to source contaminants such as waste disposal systems, toilets, and any animal being reared in the vicinity. Roof-harvested rainwater is usually collected from the roof while free-fall rainwater is collected into plastic/storage tanks directly from the sky (without the roof). Normally, free-fall rainwater will be free of risks associated with contamination sources from the roof; however in practical terms, only a small quantity of water can be collected in this way.

2.3.1 Potential Contaminants Associated with Roof-Harvested Rainwater

The quality of any source of water depends on the impact of human activities and natural processes on the composition of the water. Figure 2.2 illustrates a typical pattern in which potential contaminants find their way into the rainwater storage vessel. Firstly, as the raindrops are formed and fall from the cloud, they absorb atmospheric pollutants such as dusts and industrial and urban traffic emissions (which may include gases and wind-blown particulate from the emissions). These absorbed pollutants in the rain later drop on the roof, which may absorb more gases or other wind-blown particulate matters and animal faeces (such as lizard, rat, bird faeces etc.). This rainwater can subsequently be collected using gutter pipes or collecting tanks and can then be transferred into

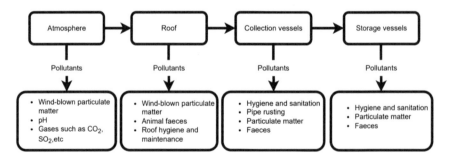

Figure 2.2 Possible associated contaminants for typical roof-harvested rainwater

storage vessels, and the condition of these tanks may have an impact on the quality of the collected rainwater (EnHealth, 2004). The contaminants in rainwater collected from a rooftop are likely to differ across national/local regions depending on specific local environmental characteristics, such as how paved an area is and how close a building is to trees or wind sources.

2.3.2 The Concept of First Flush

The concept of first flush originated in urban storm water and sewer research but is now being used in rooftop rainwater harvesting. *First flush* is the term used to describe the first wave of the dirtiest rainwater to emerge from a roof (Doyle, 2006). Dead insects, leaves, animal faeces, dust, and other particle matter can be accumulated on the roof after a prolonged period of dry antecedent days. So as the rain starts, the particulate matter on the roof is washed off (i.e., by first or subsequent waves), thereby cleaning the roof. By keeping the initial filthy water from entering the tank and allowing cleaner water to fill the tank, the first flush diversion's principal goal is to increase the overall quality of the rainwater that is collected (Church, 2001; Ntale and Moses, 2003; Martinson and Thomas, 2005). Doyle (2006) demonstrated the first flush quantity for rainwater harvesting in Rwanda and suggested that at least the first millimetre of runoff should be diverted from the roof after three consecutive days of precipitation. Martinson and Thomas (2005) found that 85% of the incoming contaminant load can be eliminated if first flush is properly executed. These studies established that while initial flushing has a positive effect on rainwater quality, it does not totally purify the water.

2.4 MICROBIAL AND PHYSICOCHEMICAL ANALYSIS OF RAINWATER

The quality of harvested rainwater in storage tanks has been the subject of numerous studies, mostly in developed regions of the world (Ahmed et al., 2010; Ward et al., 2010; Schets et al., 2010, Gikas and Tsihrintzis, 2012; Ahmed et al., 2012). There are a small number of studies on the quality of rainwater captured in Nigeria, and some of these studies are listed in Table 2.5. In those studies, most of the investigated physicochemical parameters, except for pH, were found to be within the WHO guideline limits, while microbial contaminants were commonly found in the harvested rainwater.

Acid rain is a phenomenon that primarily happens when airborne water molecules react with released gases (such as sulphur dioxide and nitrogen oxide) and contributes to the rainwater's lowered pH. This lowered pH negatively impacts infrastructure, vegetation, and aquatic creatures (Eletta and Oyeyipo, 2008). Additionally, the studies presented in Table 2.5 demonstrated that rainwater quality varies across different locations, indicating that local environmental factors may be one of the main causes for these

Table 2.5 Research on the quality of harvested rainwater in Nigeria

S/N	Authors	Salient findings	Parameters analysed	Samples collected	Nigeria
1	Abdus-Salam et al. (2011)	All parameters were within permissible limit except for pH	pH values, conductivity and 10 chemical parameters were analysed	Collector bucket	Ilorin, Kwara state
2	Okpoko et al. (2013)	All parameters were within permissible limit except for pH	4 physical parameters, 6 chemical parameters and microbial parameters were analysed	Storage tanks	Aguata-Akwa, Anambra state
3	Achadu et al. (2013)	The faecal coliform counts were above WHO guideline limits while the trace and heavy metal levels were relatively within the WHO limit except for copper and Iron levels	Electrical conductivity, pH, Total dissolved solids, acidity, trace and heavy metals and faecal coliform count were analysed	Storage tanks	Wukari, Taraba state
4	Chukwuma et al. (2012)	All parameters were within permissible limit except for pH	29 physicochemical parameters and 3 microbial parameters	Roof collected	Oko, Anambra state
5	Olaoye and Olaniyan (2012)	The faecal coliform counts were above WHO guideline limits while the physicochemical parameters were within WHO guideline limits	pH, total hardness, aluminium, copper, nitrate and sulphate concentrations were analysed.	Roof collected	Ogbomosho, Oyo state
6	Olowoyo (2011)	With the exception of pH and TSS, all others conformed to the safe limit	15 physicochemical parameters were analysed	Roof collected	Warri, Delta state

(Continued)

Table 2.5 (Continued)

S/N	Authors	Salient findings	Parameters analysed	Samples collected	Nigeria
7	Efe (2010)	With the exception of pH, turbidity, TSS, Pb, and NO_3, all others conformed to the safe limit	18 physicochemical parameters were analysed	Rain gauge	Niger-Delta
8	Olobaniyi and Efe (2007)	No coliform bacteria were found in rainwater, while coliform bacteria was found in the ground water	8 physicochemical parameters and coliform bacteria were analysed	Roof collected and groundwater	Warri, Delta state
9	Efe (2006)	With the exception of pH, Fe, TSS, and colour, all others conformed to the safe limit	23 physicochemical and coliform count	Roof collected	Delta state
10	Olabanji and Adeniyi (2005)	Al > Cr > Fe > Zn > Pb > Mn > Ni > Cu > Cd	The 9 metals in different roofs	Roof collected	Ife, Osun state
11	Adeniyi and Olabanji (2005)	Very few physicochemical parameters were beyond the safe limit and bacteria counts were higher at the start and end of rainy season	21 physicochemical parameters and bacteria counts were analysed from different roofs	Roof collected	Ife, Osun state
12	Uba and Aghogho (2000)	Almost all the physicochemical parameters were within limits, but the presence of microbes were quite high	28 physicochemical parameters and 10 microbiological parameters were analysed	Roof collected	Port-Harcourt, Rivers state

variations, such as the rate of rooftop deposition of particulate matter and its contamination with pathogens. The number of unpaved roads and non-vegetated areas is also thought to have an impact on these variables. These studies focused on the general quality of rainwater and did not examine the mechanisms that could increase the quality of rainwater coming from the storage container; this book will investigate this issue and will potentially answer the relative research questions.

2.5 ATMOSPHERIC DEPOSITION OF PARTICULATE MATTER

The main gases that make up air are oxygen and nitrogen. Other gases include ozone, carbon dioxide, methane, hydrogen, argon, and helium, which are only present in trace amounts. The atmosphere contains some naturally occurring substances, such as gases and dust particles that have been re-suspended by wind from the earth's surface. Nevertheless, anthropogenic activities have resulted in the occurrence of other compounds that are classified as pollutants because they may be harmful to human health. Heavy metals, carbon monoxide, ozone, particles smaller than 2.5 microns, particle-associated components of combustion exhaust – including cancer-causing polycyclic aromatic hydrocarbons – and sulphur dioxide are among the air pollutants that are typically dangerous to human (EPA, 2001; Stolzenbach, 2006). These compounds can exist in the atmosphere as liquid molecules, gas molecules, or aerosolised solid particles. These aerosols range in size from 0.001 to 100 microns (Stolzenbach, 2006).

Dry and wet deposition are the two main methods by which gases and particulate particles can deposit (Figure 2.3). By definition, the process by which gases and aerosols are integrated into cloud droplets, either by creating cloud condensation nuclei that being incorporated into cloud droplets, or being scavenged as the droplets fall to the ground, is known as wet deposition (Seinfeld and Pandis 1998; Izquierdo and Avila, 2012). Wet deposition occurs when raindrops carry gas and particulate matter molecules with them as they descend (EPA, 2001; Stolzenbach, 2006; Izquierdo and Avila, 2012). Snow, mist, and rain are all examples of ways that wet deposition can reach the ground or rooftops (Chantara and Chunsuk, 2008). The wet deposition of nitrogen and sulphur compounds, which cause acid rain, is one of the biggest threats to the environment. The most basic measuring tools for wet deposition comprise a receiver, such as a funnel, coupled to a gallon, bottle, bucket, or tray. A typical receiver features a programmed cover that slips away from the receiver to open when it is raining and let debris out when it is not raining (Lovett, 1994; EPA, 2001; Izquierdo and Avila, 2012). Almost any pollutant, as well as some contaminant isotopes, can be measured using this technique.

The combination of molecular diffusion, impaction, and gravity settling constitutes dry deposition (EPA, 2001; Stolzenbach, 2006). Turbulent transfer

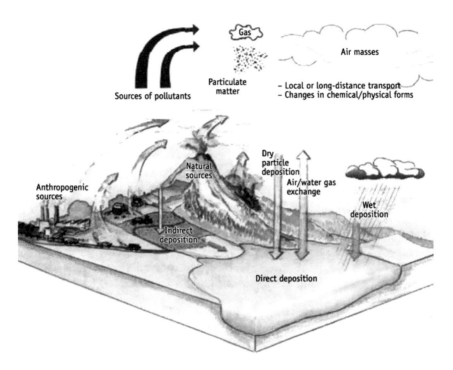

Figure 2.3 Atmospheric deposition processes (Source: EPA, 2001; Stolzenbach, 2006)

and gravitational settling cause the dry deposition of gases and particles over water and land surfaces (Lovett 1994). Wet deposition has been described as the most efficient scavenging factor for removing particulate as well as organic and inorganic gaseous pollutants from the atmosphere (Arsene et al., 2007; Al-Khashman 2009;, Prathibha et al., 2010; Izquierdo and Avila 2012). However, dry deposition may have a higher contribution in arid environments as precipitation is limited (Dolske and Gatz 1985; Guerzoni et al., 1999; Aas et al., 2009; Izquierdo and Avila, 2012). Dry deposition can be measured using a surrogate surface method that involves collecting gases and dry particles on a surface over time. A collector can also be used to measure the amount of dry deposition on a given area. The dry deposition in an area can either be obtained by measuring the amount of dry particles and gases in the air or by multiplying the dry deposition velocity of particles (m/year) with the total (gas and particle) ambient concentration (g/m^3) (Lovett, 1994; EPA, 2001; Izquierdo and Avila, 2012).

Wet deposition is the most significant deposition mechanism in regions with significant annual precipitation, while dry deposition methods are more likely to control atmospheric deposition in semi-arid regions (Stolzenbach, 2006; Prathibha et al., 2010; Izquierdo and Avila, 2012; Qiao et al.,

2015a,b). Gravitational settling of particles between 10 and 100 microns makes up the largest percentage of the dry deposition that forms most efficiently. Atmospheric deposition has largely been disregarded when analysing the impact of air pollution on human health because the dry and wet deposition rates for very small particles and most gases are low (Stolzenbach, 2006; Prathibha et al., 2010). Acid rain is the most well-known source of atmospheric deposition, and some of the most significant lakes, rivers, and water bodies have experienced considerable pollution because of atmospheric deposition. Therefore, how deposition builds up on rooftops and how subsequent microbial activity affects the quality of rainwater are discussed in this book.

Between June and August 2010, Qiao et al. (2015a) carried out a study in the Jiuzhaigou National Nature Reserve of China to identify the deposition contributions made by the various location sources as well as to pinpoint the origin sectors that were responsible for the fluxes of deposition. The wet and dry depositions of certain ions were also examined in that literature. The studied ions include ammonium (NH_4^+), nitrate (NO_3^-), and sulphate (SO_4^{2-}). Results from the three-month study showed that wet deposition fluxes of NH_4^+, NO_3^-, and SO_4^{2-} produced 0.56 kg N ha^{-1}, 0.39 kg N ha^{-1}, and 2.42 kg S(N) ha^{-1}, respectively, whereas the dry deposition fluxes of the NH_4^+, NO_3^-, and SO_4^{2-} produced 0.21 kg N ha^{-1}, 0.08 kg N ha^{-1}, and 0.07 kg S ha^{-1}, respectively (Qiao et al., 2015a). After examining how different industries and activities, such as the fertiliser, biogenic, power, manure management, and transportation industries, as well as domestic and open burning, contributed to the deposition fluxes in this area, it was found that the fertiliser and manure management industries were responsible for 87% of the ammonium ions. Industrial sources were responsible for 86% of the sulphate ions and 51% of the nitrate ions, while domestic sources produced much lower emissions. It was determined that the majority of the emitted ions were due to long-distance transport rather than local emission because more than 90% of the total deposition fluxes came from distant places (Qiao et al., 2015b).

In a rural location in the north-eastern region of Spain, another study was conducted to compare the methodologies employed to assess the atmospheric deposition of certain particles and ions (Izquierdo and Avila, 2012). These samples were collected on a weekly basis between February 2009 and July 2010, where NO^{3-}, Mg^{2+}, SO_4^{2-}, K^+, Cl^-, Ca^{2+}, Na^+, and NH_4^+ were analysed. The findings illustrated that wet deposition removed 74% of the overall deposition, which was greater than dry deposition. According to their investigations, bulk deposition consisting of a combination of dry and wet depositions was the most effective of the three depositions at capturing pollutants, and this is because bulk deposition is carried out throughout both rainy and dry periods (Izquierdo and Avila, 2012). In another study by Pan and Wang (2015), the wet deposition fluxes were, however, found to be significantly lower than the dry deposition fluxes for the majority of the tested elements that measured the dry and wet depositions in Northern China. From these studies, it can be inferred that the deposition of these fluxes is typically influenced by local environmental factors, including the local intensity of the precipitation, the size distribution of the tested elements or particles, the season (rainy, dry, winter, summer, autumn, and spring), the local wind patterns, and land use activities in the area. These local environmental factors influence the wet, dry, and bulk/total depositions.

2.6 TRANSPORT AND SEDIMENTATION OF PARTICULATE MATTER

The processes involved in the atmospheric deposition of particulate matter have been described in the previous section. This section will explore the transportation and sedimentation of these particulate matters from atmosphere to storage tank of the roof-harvested rainwater. In general, atmospheric particulate matter can be defined as a mixture of materials of various sizes; its chemical makeup may change over time and from one place to another depending on the environment and the types of discharge sources (Sprovieri et al., 2008). Physically, particulate matter is a combination of extremely microscopic particles and liquid droplets suspended in the earth's atmosphere. There are long-lasting issues for the biological and geochemical cycles of trace elements related to the atmospheric transport of suspended particulate matter (Marcazzan and Persico, 1996; EPA, 2001; Koc-ak et al., 2004a, b; Stolzenbach, 2006; Sprovieri et al., 2008). Researchers have studied how airborne micro-organisms interact with atmospheric processes and have found that the atmosphere contains some micro-organisms, including bacteria and fungi (Pouleur et al., 1992; Pruppacher and Klett, 1997; Morris et al., 2011; Goncalves et al., 2012). Research by Goncalves et al. (2012) revealed that microbes like *P. syringae* can be found in the atmosphere and that they are active at high temperature and warm weather.

Several studies have evaluated the possibility of using rooftops as conduit or source for contaminated runoff into the ecosystem. While some studies claimed that the contaminants from rooftops were extremely dangerous to aquatic organisms, others also demonstrated that the quality of roof-harvested rainwater exceeded drinking water quality guidelines (Yaziz et al., 1989; Good, 1993; Thomas and Greene, 1993; Uba and Aghogho, 2000; Van Metre and Mahler, 2003). Adeniyi and Olabanji (2005) compared the quality of roof-harvested rainwater and free-fall harvested rainwater in a western state of Nigeria. The results showed that the total suspended solid (TSS) values for both the free-fall and roof-harvested rainwater ranged from 6.7 to 25.6 mg/L and 16.2 to 112.7 mg/L respectively; while the turbidity values for both ranged from 1.7 to 9.5 NTU and 2.2 to 38.3 NTU, respectively. The result showed that the quality of the free-fall rainwater was better than that of the roof-harvested rainwater, and this difference in the water quality was attributed to the roof being a pathway for contamination. The study also noted that the quality of roof-harvested rainwater can be impacted by the roof's composition, age, and location. Metallic roofs typically produce rainwater with superior microbiological quality compared to other types of roofs. This is linked to the dry heat that is typical of metallic roofs, which efficiently eliminates numerous bacteria under bright sunlight, especially in tropical nations (Yaziz et al., 1989; Vasudevan et al., 2001; Adeniyi and Olabanji, 2005; Meera and Ahammed, 2006; Mendez et al., 2011; Eruola et al., 2012). The position of the roof in relation to its proximity to the source of contamination is another element that affects the standard of roof runoff. For instance, a roof that is near a tree where birds live is more likely to result in lower quality roof-harvested rainwater as compared to a roof that is not near a tree (Forster, 1996; Meera and Ahammed, 2006). On the other hand, the correlation between various roof materials and the quality of roof-harvested rainwater showed that water quality declined as roof materials aged (Adeniyi and Olabanji, 2005; Eletta and Oyeyipo, 2008). This was determined by the fact that roofing material wear-off proved to be active as the roof ages, increasing pollutant counts in the collected rainwater.

Van Metre and Mahler (2003) investigated how contaminants in urban streams were affected by particles from rooftops. This was accomplished by evaluating the contribution and yield of rooftop runoff to the watershed from two different types of roof materials. Findings revealed that rooftop runoff contributed 45%, 46%, and 55% of the total particle-bound watershed loads of lead, mercury, and zinc, respectively. Characklis et al. (2005) examined the microbial partitioning of settle-able particles in storm water and found that 20%–35% of bacterial organisms were associated with these particles in background samples and 30%–55% in storm samples, while 20%–60% of the total coliphage were associated with the particles during the stormy period. Their findings also reported that the highest average amount of particle attachment was found in the spores of Clostridium perfringens, with bacteria count values ranging from 50% to 70%. They have also found that the type of microbe and the partitioning behaviour typically differ between wet and dry climatic conditions when it comes to the attachment of bacteria to settle-able particles. It was discovered that rain events preferentially transfer atmospheric nitrate and acidity, as seen in a study investigating the long-range transport and chemical composition of wet air deposited samples in Opme, France (Bertrand et al., 2009). That study, along with others, brought attention to the importance of investigating the effects of atmospheric deposition of particulate matter to roof-harvested rainwater.

Different sampling frequencies have been employed to measure deposition in various locations and nations. The purpose of the utilised data typically determines the chosen frequency of sampling. These frequencies include sampling on a 12-hour, daily, weekly, monthly, or annual basis. Sample frequencies, such as weekly, monthly, and yearly depositions, are often sufficient to estimate deposition over a longer period. Shorter sampling frequencies, such as 12-hour and daily, are mostly used to determine individual emission sources and depositional processes (EPA, 2001; Izquierdo and Avila, 2012). Estimating the amount of particle matter that has been deposited in the studied area of the Lagos state utilised in this book is one of the goals of this work. To achieve this objective, it was necessary to determine daily, weekly, and monthly depositions. Thus, data from these sampling intervals was utilised to define the sources of the pollutant concentrations and wind speed. Additionally, it was crucial to look at how the wind speed and other factors affect both the long and short sampling depositions due to the frequent changes in wind speed and direction over time.

2.7 QUANTITATIVE MICROBIAL RISK ASSESSMENT

The Quantitative Microbial Risk Assessment (QMRA) is a standard technique that is used to assess the likelihood of adverse impact on human health of exposure to a specific risk, and to inform opinions and choices regarding risk management and mitigation. The QMRA is commonly adjusted to highlight potential dangers to humans from exposure to or ingestion of pathogens, where this may include fate and transport models for multiple media, such as the source zone (faecal release), air, soil/land surface, surface water, vadose zone, and aquifer (Whelan et al., 2014). Hazard identification, dosage response, exposure assessment, and risk characterisation and management are the four categories of information used by the QMRA to characterise potential dangers to human health (Haas et al., 1999; Hunter et al., 2003; Whelan et al., 2014).

Hazard Identification: This incorporates a wide variety of information about infectious agents and covers a broad range of information about the pathogen, the detrimental effects on the host from infection, and the pathogen in general. This is often performed by compiling research articles reporting on the presence or absence of the target diseases (Ahmed et al., 2010). This phase of the QMRA process is an early assessment of data that will be scrutinised further in subsequent phases. It mostly consists of a qualitative assessment of the risk (Lammerding and Fazil, 2000). Determining whether there is sufficient evidence to show that a substance, such as *E. coli*, is the cause of a negative health consequence, i.e., diarrhoea, is one of the phase's key goals (Lammerding and Fazil, 2000; Ahmed et al., 2010).

Exposure Assessment: The simplest definition of exposure is the amount of the pathogen that a person consumes. To determine the likelihood of infection, this number is entered into the dose-response model within QMRA. The exposure evaluation also identifies the routes by which a microbe can infect people and spread, i.e., via consuming rainwater. The exposure assessment also calculates the magnitude and duration of exposure by each pathway, as well as the estimated exposure rates and affected population categories (Petterson et al., 2007; Whelan et al., 2014). The considered pathway included the micro-organisms that were counted both before and after the application of various household treatment techniques.

Dose Response Assessment: A target pathogen's known dose (number of micro-organisms) is used in this phase to evaluate the risk of a response. For a given target pathogen, transmission pathway, and host, the dose response models are statistical equations that specify the dose–response connection. Several human and animal research data sets were used to create the statistical model for predicting dose response (Petterson et al., 2007; Whelan et al., 2014). The amount of *E. coli* consumed both before and after the various HHTTs are measured, and the results are used to calculate the population's likelihood of contracting a disease.

Risk Characterisation: To estimate harmfulness, information on the dose received (from the exposure assessment) is combined with risk information related to various doses (from the dose response assessment). Risk evaluation will include combining the data from the three processes mentioned above (hazard identification, dosage response, and exposure assessment) into a single statistical model. With the aid of this statistical model, the likelihood of events like infection, illness, or death can be calculated. For exposure, dosage, and hazard estimation, the three phases will offer a range of values rather than just one. The Monte Carlo analysis is used in this situation, and the outcomes show the whole spectrum of potential risks, including average and worst-case possibilities. These are also the hazards that policymakers consider when making decisions and that scientists study to determine where we need to conduct additional experiments to obtain better data and information (Petterson et al., 2007; Whelan et al., 2014).

Risk Management: Risk management aims to reduce and minimise risks and any unfavourable effects they may have. Different approaches to risk management are possible, and they are most successful when they are based on risk characterisation (Petterson et al., 2007; Whelan et al., 2014). Risk management in this book involves an examination of the usage of various household treatment techniques by the locals who drink rainwater and to advise on the best treatment options. Additionally, it instructs the rainwater harvester to do a first flush due to rooftop depositions, which may be a source of microbial contamination, especially in the areas that have high rooftop deposition rates.

The QMRA was piloted in a semi-urban area of Argentina to assess the relative health risk of untreated water supplies that are used for drinkable purposes (Rodriguez-Alvarez et al., 2015). *Giardia, E. coli,* and *Pseudomonas aeruginosa* (*P. aeruginosa*) were the three target pathogens that were examined in that study. The findings revealed that *E. coli* risks were nearly in line with the WHO guideline of 1:10,000 for untreated water sources, while the risks from *Giardia* and *P. aeruginosa* were well above the acceptable guidelines in nearly all cases (Rodriguez-Alvarez et al., 2015). In a rural Mexican community, Balderrama-Carmona et al. (2015) determined the health risks of *Cryptosporidium* and *Giardia* associated with drinking well water using the QMRA technique. The results show that both protozoa were present in every well-water sample. The QMRA results also showed that the average yearly risks for *Giardiasis* and *Cryptosporidium* were 100% and 99%, respectively. The health risk presented by the urban drinking water system in the Indian state of Karnataka was estimated by George et al. (2015). The QMRA findings revealed that while the risk from *E. coli* and *Campylobacter* was higher than the WHO guideline value, the overall annual risk from rotavirus was lower. According to the QMRA findings from the reviewed literature, the majority of the environmental water was assessed to be dangerous for consumption, and those who drank water from these sources were more likely to develop gastrointestinal disorders. The likelihood of infection by various target diseases from various water sources or supplies has been evaluated by several authors using the QMRA technique in various parts of the world. It is also clear that monitoring and evaluating the quality of the water is crucial to be able to suggest the best treatment method for the water sources.

2.8 SUMMARY

This chapter examined the movement of atmospheric deposition and the quality of collected rainwater in various regions of the world, demonstrating that rainwater's quality and deposition rates vary depending on location. This might be due to regional environmental issues. Additionally, the QMRA was discussed, and numerous contaminants connected to roof-harvested rainwater were highlighted, along with how a conventional rainwater collecting system operates. A thorough description of the researched regions and the requirement for drinking water analysis were explored.

Chapter 3

Rainwater Technologies: Equipment and Analysis

ABSTRACT

Both state-of-the-art and standard techniques employed in determining the deposition rate and quality of rainwater are described in this chapter. State-of-the-art techniques described include the measurement of net deposition, roof runoff deposit, fractionation of the sample by serial filtration, and roof surface wind speed. Standard techniques described include the measurement of total deposition, total suspended solids (TSS), pH, turbidity, the Colilert method, rainfall, and wind speed. The equipment used for both types of technique was tested in the laboratory and calibrated to APHA, AWWA, and WEF (2005) standards before the experiments.

3.1 STATE-OF-THE-ART EXPERIMENTAL METHODS AND EQUIPMENT

Previous writers have determined their studied depositions either by wet, dry, or bulk/total deposition (see Sections 3.1.1 and 3.1.2). This study determined net deposition either as a roof runoff deposition in the rainy season or as net deposition in the dry season.

3.1.1 Net Deposition Rate

The net deposition experiment determines the mass of particulate matter that is deposited on a rooftop by direct measurement. The described net deposition refers to the equilibrium of deposition and re-suspended solids, i.e., at Equation 3.1. This was done mainly in the dry season, as limited or no rain events occurred during this period. To investigate the deposition rate in depth, experiments conducted by the authors have also been used in this book to compare them with the referred knowledge from literature. The roof examined in the proposed experiment was exposed to the atmosphere for the duration of 3, 7, 14, 28, and 56 days in alignment with the objectives set up and discussed in Chapter 1.

DOI: 10.1201/9781003392576-3

$$Net\ deposits = Total\ deposits - suspended\ deposits\ by\ wind \qquad (3.1)$$

Galvanised corrugated sheets were fixed to the roof for the above-mentioned predetermined amount of time, and then the accumulated deposit was measured. Before being fixed, the galvanised corrugated sheets were cleaned with sterile water to avoid external contamination. No *E. coli* or total coliform were found when the sterile water was tested for bacteria (0 MPN/100 mL). The dimensions of the metal tile are determined by its function for the deposition rate and limit of detection of the gravimetric analysis, where a 80cm × 60cm tile was used.

The tiles were securely fixed to the roof to prevent them from falling and were taken off carefully after reaching the predetermined amount of time for analysis. After the tile was removed, it was put in a box with a secured lid (to prevent the deposits from being blown away) and carefully transported to a location where it was rinsed off for analysis. It was ensured that all the deposits were removed using mechanical agitation. After washing the sheets, the water was collected and taken to the lab in an icebox. At that point, it was divided up if microbiological or other tests were required.

The mass retained on the filter was calculated after a known volume of the water sample was filtered through 1.5 µm of filter paper, according to the standard TSS retrieving technique. In addition, the 2 L of water samples was filtered through the 1.5 µm filter paper for the net deposition tests in this experiment. The net deposition (g/m^2/day) could then be calculated by Equations 3.2 and 3.3.

$$Total\ Solids\ (g) = (Residue + Dry\ Filter)\ (g) - Dry\ Filter\ (g) \qquad (3.2)$$

$$Net\ Deposition(g/m^2.day)$$
$$= \frac{Total\ Solids(g)}{\left(Corrugated\ Sheet\ Area\,(m2) \times Period\ of\ Exposure\,(days)\right)} \qquad (3.3)$$

In addition to providing solutions to the research questions, this experimental approach was chosen because it was affordable, effective, and simple.

3.1.2 Roof Net Deposit Runoff

The roof runoff experimental method determined the volume of particulate matter removed from a gutter catchment area in each rainfall event. The rainfall in the rainy season was monitored using a standard method of experiment (see Section 3.1.5). Before the experiment, sterilised water was used to wash the plastic bottles and storage tanks to ensure there was no *E. coli* or total coliform and the roof was free from contamination before the rain and collection. A gutter was used for capturing rainwater runoff from a specified

Figure 3.1 0.5 m × 20 cm × 10 cm gutter used to measure the roof runoff deposit

property's roof into a storage tank (Figure 3.1). The dimensions of the gutter and storage container for the subject property were 4 m × 2.5 m and 255 L, respectively.

After rainwater was collected using a 20 cm × 10 cm gutter, it was stored in a 255 L (diameter 60 cm, height 90 cm) storage vessel that was positioned underneath the roof. At the end of the rainfall event, the thoroughly mixed water samples were transferred from the storage tank into a prewashed plastic container. Equations 3.4 and 3.5 utilise the above collected information to calculate the runoff deposition (g/m²).

Total Solids Runoff (g) = (Residue + Dry Filter) (g) – Dry Filter (g)   (3.4)

$$\text{Roof Runoff Deposition}\left(g / m^2.\text{day}\right)$$
$$= \frac{\text{Total Solids (g)} * \text{Volume of Rainwater in Tank}}{\left(\text{Corrugated Sheet Area (m2)} \times \text{Period of Exposure (days)} \times \text{Volume of Sample}\right)}$$

(3.5)

Every effort was made to mix the harvested rainwater in the tank properly in order to obtain a representative sample. When the volume of the collected rainwater was less than 10 L, the storage tank was gently shaken; when it was more than 10 L, it was thoroughly mixed for 3–5 minutes using a

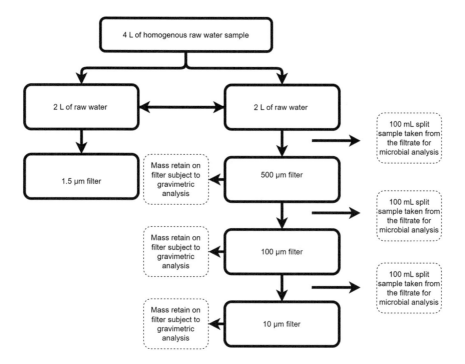

Figure 3.2 Experimental procedures for serial filtration

clean spatula. The built-in gutter was cleaned with sterile water before being fastened to the roof, and the storage tanks were cleaned with sterile water both before and after each rainfall event to prevent record of additional contamination.

3.1.3 Fractionation of Samples by Serial Filtration

The fractionation technique creates sub-samples of suspended particulate matter in a specified size range for further gravimetric and/or microbiological investigation. Figure 3.2 presents a schematic representation of the experiment's concept. For the experimental method to be successful, the volume of processed water should show that

- It is sufficient to yield at least 100 mL water at each stage for further microbiological testing.
- The mass retained on each filter must be greater than the limit of detection of the gravimetric method.

Therefore, a sample of 500 mL needed to be collected, which is divided into 5 equal portions to produce 100 mL of sample that leaves a measurable deposit on each filter. The testing was done using a 4000 mL water sample during the field work, where 2000 mL of the raw water from the original sample was taken out for additional analysis after being gently shaken to create a homogenous solution. A filter with pore size of 1.5 µm was used to filter the raw water after it had undergone microbial analysis. Before and after the water sample was passed through the 1.5 µm pore size filter, its weight was measured. This is important to determine the deposit's weight and the number of bacteria present in the sample for further analysis.

After that, the 500 µm pore size mesh filter was used to filter the remaining 2000 mL of the water sample. To calculate the mass of the retained deposit, the weight of the filter with a 500 µm pore size was recorded both before and after the water sample filtered through it. Subsequently, 100 mL of the filtrate was separated for microbiological analysis. The remaining 500 µm filtrate was then put through a 100 µm filter, and the procedure above was repeated. Then, the filtrate (from the 500 µm and 100 µm filters) was put through a filter with a 10 µm pore size. Before and after the water samples were passed through the filter, their weight was determined. Lastly, each filtrate from every single filter was then taken for microbiological analysis.

In total, this experimental procedure yielded the following samples:

1. Raw sample (1.5 µm filter)
2. Passing 500 µm filter
3. Passing 100 µm filter
4. Passing 10 µm filter

The 1.5µm filter gave the total solids present in the raw samples (by assuming the solids smaller than 1.5µm are insignificant and nearly weightless), while other filter sizes gave the range of the different sizes of the particulate matter.

3.1.4 Roof Surface Wind Speed with Hot-Wire Anemometer

This section details the instrument used to monitor wind speed at the roof surface. This wind speed may be substantially different from that in the general and public vicinity. A hot-wire anemometer (ModernDevice, 2017) was set up on the surface of the roof to investigate these variations and their effect on rooftop deposition. Throughout the recording, this device was mounted 1 metre away from the roof in order to measure the wind speed whilst avoiding significant variations due to the impact from the roof.

The Arduino Uno rev 3 was used to power and manage the hot-wire anemometer, and an Adafruit SD card shield was used to capture data. The manufacturer-generated calibration curve used by the Arduino script is dependent on the microcontroller board's internal 5 volts regulator. A wind sensor functions as a thermal anemometer, using a traditional wind speed measurement

Figure 3.3 10-inch funnel attached to the 5000 mL plastic container for measuring total deposition

technique. The hot-wire method involves heating the component to a constant temperature and then observing how much electrical power is required to keep it at that temperature while wind varies.

3.2 STANDARD METHODS AND EQUIPMENT

3.2.1 Total Atmospheric Deposition

The total atmospheric deposition method determines the mass of particulate matter that deposits from the air through a flat-plane surface, i.e., a roof. The method involved exposing the experimental vessel for a set time in the environment. The experimental vessel is a 5000 mL transparent plastic container fitted to a 10-inch funnel (such as the simple setting presented in Figure 3.3). The pre-sterilised container is partially filled with sterilised water. Particulate matter deposited on the funnel was collected in a container, which was pre-sterilised and partially filled with sterilised water. The funnels were also thoroughly rinsed with sterilised water before the beginning of every sampling period to prevent external contamination that would corrupt the measured data.

The 5000 mL plastic container was exposed for 7 days to obtain the weekly total deposition. At the end of that period, sterilised water was used to rinse the particulate matter deposited on the funnel into the container. This would subsequently result in the water and associated solid depositions being withdrawn completely. In this instance, the water sample was split into portions for microbiological or other tests. The procedure shown in Section 3.1.1 was used to calculate the solids and total depositions over the 7-day period.

A sampler was used to measure the atmospheric deposition in a rural Mediterranean site in the north-eastern part of Spain by Izquierdo and Avila (2012). The measurement method for total wet and dry deposition was presented separately. In the determination of total deposition, the funnel in the sampler collected deposition, where the recorded total solids over the targeted period was divided up based on the area of the funnel. The wet deposition accumulated during a rain event, while dry deposition gathered when there was no rain event (dry conditions). The collector was prepared with two polyethylene buckets (290 mm inner diameter) and a covering lid, which was shut. It was removed at the beginning of each rain event (Izquierdo and Avila, 2012).

The total deposition used for measurement in this book was collected and measured using ta similar method to that used by Izquierdo and Avila (2012) in both rainy and dry seasons. In comparison, the net deposition (shown by the method at Section 3.1.1) is influenced by the resuspension of dust in both seasons. The difference between the total and net depositions is the result of external factors, such as wind. Wind has limited or no effect on the measured total deposit from the funnel experiments, while the net deposition from the tile area is significantly affected by wind.

3.2.2 Microbial Testing Procedure

The *E. coli* and total coliform concentrations were enumerated using the standard Colilert-2000® procedure (IDEXX, 2023). The Colilert technique was executed in accordance with the manufacturer's guidelines. In its procedures, 100 mL of the sample was added to IDEXX's dehydrated media in the supplied sterile jars. The samples were shaken three to four times over 6 minutes to dissolve the media. The contents in the jars were then emptied into sterile quanti-trays and heat sealed with sealer. The quanti-trays were incubated in accordance with the manufacturer's guidelines at 35°C for 18 h by Colilert-18 and for 24 h by Colilert-24. Following incubation, the quanti-trays were compared with the supplied comparator. Then, the yellow wells were enumerated and, by using the MPN table, the number of coliforms were recorded. After that, the quanti-trays were placed in the fluorescing wells (366nm), and the number of *E. coli* cells were also counted and recorded. The MPN table used in the experiment is shown in Appendix F. All experiments were performed in duplicate to allow the same sample to be analysed twice. The microbial dilutions were done in ratio of 1 in 10 (i.e., for every 100 mL of the sampled water, 10 mL is the raw sample, while the remaining

90 mL is sterilised water). The Colilert method was used to detect *E. coli* and total coliform because it is relatively easy to use, rapid, and relatively accurate (ISO-1908-1, 2000; ISO-17994, 2004; Mannapperuma et al., 2011; IDEXX Laboratories, 2016).

It is worth noting that there are several available methods for analysing and detecting total coliforms and *E. coli* bacteria in water samples. These methods include multiple tube fermentation, membrane filtration, m-Coli-Blue24, and the above-described Colilert methods (Mannapperuma et al., 2011). After comparing the different methods used to enumerate and detect total coliform and *E. coli*, Mannapperuma et al. (2011) found that alternative methods (i.e., Membrane Filtration, m-ColiBlue24, and Colilert methods) were in a good sensitivity range and were efficient compared to the multiple tube fermentation method. They also recommended that the Colilert and m-ColiBlue24 methods are very suitable for analysing drinking and surface water samples in tropical countries as they gave superior performance. However, they are more expensive compared to the membrane filtration and multiple tube fermentation methods (Mannapperuma et al., 2011).

3.2.3 Total Suspended Solids

The total suspended solids (TSS) were determined using a vacuum filtration device. The steps used to determine TSS in the water samples are as described by the standard method in APHA, AWWA, and WEF (2005). The TSS was determined using the following calculation.

$$\text{TSS}\,(\text{mg/L}) = \frac{(\text{Residue} + \text{dry filter}) - \text{dry filter}\;(\text{mg})}{\text{sample filtered}\;(\text{L})} \qquad (3.6)$$

3.2.4 Wind Speed

The wind speed was determined using a specific data logger, i.e., the Omega OM-CP-WIND101A-KIT Series. The OM-CP-WIND101A is a three-cup anemometer, manufactured by the Omega Company, and comes with a data logger in a weatherproof enclosure (Omega, 2017). It was attached to a 3000 mm pole, which was mounted in the area of analysis. The three-cup anemometer was placed near to the hot-wire anemometer in the proposed study area within this book (as showing by Figure 4.3 in Chapter 4), and their recorded data were compared to assess their differences (detailed comparison is presented in Section 4.2). The Omega OM-CP-WIND101A-KIT Series was tested in the field and was calibrated using the manufacturer's manual guideline. Besides, it was also calibrated according to the World Meteorological Organisation's guideline when the anemometer was installed (before using for the experiments).

3.2.5 Rainfall

The rainfall intensity was obtained using a data logging rain gauge, and this was manufactured by EnviroMonitors Company. The instrument meets the World Meteorological Organization requirement, which specifies a minimum 205 mm diameter collector. Prior to using the equipment, the tipping spoon and rain funnel were cleaned with a dry cloth to remove particles and water. This rain gauge was placed in an open area on some block-works, 1000 mm from the ground level. This was done to prevent contaminants and water from splashing into the collector. This equipment was tested in the field and was calibrated using the manufacturer's manual and guidelines before the experiments.

3.2.6 Turbidity

The turbidity of the water samples was determined using the Hanna Turbidimeter, manufactured by Shanghai Xinrul Instrument and Meter Co Ltd. This equipment was tested in the laboratory and was calibrated to the set standards by APHA, AWWA, and WEF (2005) before the experiments.

3.2.7 pH

The pH of different water samples was analysed using PHS-3D pH meter, which is manufactured by ShangHai San-Xin Instrumentation Inc. This equipment was tested in the laboratory and was calibrated to APHA, AWWA, and WEF (2005) standards before the experiments.

3.2.8 Conductivity

Lastly, the conductivity was measured using an electrical conductivity meter (Palintest waterproof 800), which is manufactured by Palintest Instrument. To ensure its calibrated feature, it was tested in the laboratory and was calibrated to APHA, AWWA, and WEF (2005) standards, where all these tests and calibrations were performed before the experiments in this book took place.

3.3 SUMMARY

The standard and state-of-the-art techniques for measuring depositions and water quality parameters have been covered in this chapter. The standard experimental methods are described by APHA, AWWA, and WEF (2005) in the examination of water and wastewater, whereas the state-of-the-art experimental methods were developed or improved in the study conducted within this book. The rates of particulate matter deposition on a chosen building in the research location during dry and rainy seasons are the main discussion of the next chapter.

Chapter 4

Atmospheric Deposition Impact on Rainwater Quality

ABSTRACT

A building's roof represents a critical pathway for atmospheric sediment to mix with rainwater, and atmospherically deposited particulate matter impacts many aspects of human life. This chapter assesses how rooftop deposition varies according to the season and how variables such as rainfall intensity, wind speed, and dry antecedent day (DAD) correlate with rooftop depositions. Fractionation tests (serial filtration) were conducted to obtain the range of the particle sizes and to enumerate the number of bacteria attached to the deposits. The quality of rainwater is significantly impacted by particulate matter deposition as well as the magnitude and type of microorganisms attached to the deposit, which is further dependent on the source, hygiene, and prevalence of wind speed. In short, this study will provide the vital link between atmospheric depositions and collected rainwater quality.

4.1 INTRODUCTION

Particulate matter is made up of microscopic solid or liquid droplets that can be breathed in and cause major impacts on human health. Examples of particulate matter from natural sources include pollen, sea salt, dust, and airborne dirt. It also includes material from volcanic eruptions and particles produced from naturally occurring gaseous precursors such as sulphates (StatsNz, 2021). Particulate matter deposition from the atmosphere impacts many facets of human life, and thus it should not be disregarded. It has been described as a source of ocean nutrients, i.e., for flora and fauna, and fertiliser for forests (Martin et al., 1991; Swap et al., 1992; Pu, 2016).

Studies have evidenced that Saharan particulate matter depositions affect the earth's radiation balance, air quality, and precipitation in the sub-Saharan region (Prospero, 1999; Prospero and Lamb, 2003; and Tegen et al., 2004). Additionally, visibility can be significantly reduced by these dusts/particulate matters, particularly during the Harmattan period, which is a peak period of particulate matter deposition that primarily takes place between

DOI: 10.1201/9781003392576-4

December and January every year in the sub-Saharan region of Africa. For instance, a Kenyan airline crashed near Abidjan on January 30, 2000, after its forced landing in Lagos, Nigeria. This was a result of the Harmattan dust that engulfed the airport and ultimately caused flights to and from Lagos to be suspended (Pinker et al., 2001; Sunnu et al., 2013). Reduced activity days, health issues with the heart and lungs, and premature baby deaths are all associated problems caused by exposure to particulate matter (StatsNz, 2021). Data on ground and atmospheric particulate matter deposits in the countries impacted by Harmattan dusts is limited. Thus, determining the rates of particulate matter deposition in this region of Nigeria and its impacts are some of the set objectives for this chapter.

It has been established that Saharan particulate matter periodically enters the West African countries close to the Gulf of Guinea during the dry season, which lasts from November to March every year. Due to fewer or no rainfall events during this dry season, higher particulate matter deposition rates have been documented compared to the rainy season (e.g., McTainsh, 1980; D'Almeida, 1986; Tiessen et al., 1991; Afeti and Resch, 2000; Breuning-Madsen and Awadzi, 2005; Resch et al., 2007; Sunnu et al., 2006; Lyngsie et al., 2011; Sunnu et al., 2013). During the Harmattan period, this dry and hazy dust travels from the Bodélé Depression of the Chad Basin, which is believed to be the lowest point in Chad and is situated at the southern edge of the Sahara Desert in the north central region of the country (Brooks and Legrand, 2000; Prospero et al., 2002; Tegen et al., 2006; Todd et al., 2007; Prospero, 2011; Sunnu et al., 2013). The physical properties, flux, and transportation of the particulate matter in the rural areas of the southern region of Nigeria have rarely been assessed, while limited studies have been conducted on the periodic influx into the neighbouring countries surrounding this area. Due to this, the investigation of the range of sizes of the deposited particles and their associated bacteria in both rainy and dry seasons in a suitable studied area will be the next aim of this chapter.

The west of the Sahara includes the North Atlantic, the Caribbean, and West Africa. The transportation of dusts from the Saharan desert towards the west is more persistent and has recently caught the attention of atmospheric pollution researchers, while the transportation of dust to the east is occasional. The occasional dust movement to the east is due to the counteracting movement of the Saharan dust that prevents its eastward transport (Sunnu et al., 2013). The arrow in Figure 4.1 illustrates the typical direction of the wind which entrains dust originating in the Bodélé Depression (Faya-Largeau) across the countries that traverse towards the Gulf of Guinea. Since Nigeria shares a border with Chad, which is along the dust particles' path of travel, there is always a significant dust deposition in Nigeria (Sunnu et al., 2013). In this study, an investigated area situated in the southern part of Nigeria has been utilised to explore the environmental impact of atmospheric dusts and particulates, as it is imperative to assess their rate of deposition in this part of the country.

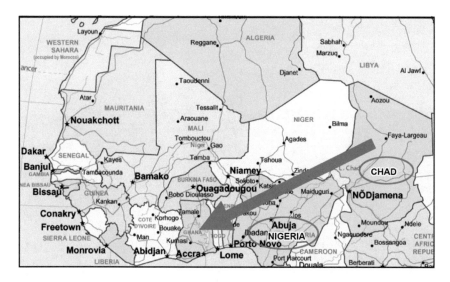

Figure 4.1 Map of West Africa showing the direction of the wind from the dust source Faya-Largeau in the Bodélé Depression through Nigeria towards other West African countries

Several factors such as soil texture, wind speed, vegetation, vegetative residue, surface roughness, soil aggregate size distribution, soil moisture, and rainfall have been identified as influencing the concentration of atmospheric dusts in different parts of the world, especially surrounding desert regions. For example, the agricultural use of the lands adjacent to the desert dust source (i.e., the Faya-Largeau in the Bodélé Depression) and the drought in the Sahel lands was significantly associated with the increase in dust source strength in the Sahara Desert (Mulitza et al., 2010; Sunnu et al., 2013). In view of the importance of these factors, this book will discuss some of them, including wind speed, vegetation, periods of dry antecedent days, and amount of rainfall. This chapter further determines the characteristics and constituent range sizes of the deposited particulate matters in this southern part of Nigeria using a selected study area.

4.2 RESEARCH RATIONALE AND STUDY AREA

To illustrate the atmospheric deposition impact on rainwater, a studied site has to be employed. The Ikorodu area of Lagos state has been chosen here (where the study area and house type are fully discussed in the follow-up paragraphs), with an emphasis on the serial filtration of the sample. The range of particulate deposits in the study area was characterised, and finally the percentage of the bacteria associated with deposited particles was also determined. This was done in both rainy and dry seasons. The primary hypothesis

Figure 4.2 Property used in study – view 1

to be tested in this chapter is the following: *Rooftop deposition will vary significantly with seasons and will be particularly high during the Harmattan period. Further, deposition will accumulate over time until an equilibrium level is reached that depends on local conditions, i.e., wind speed and the number of dry antecedent days.*

The set of experiments conducted in this section were undertaken at a property in Ikorodu, Lagos, Nigeria. A representative building (based on common building features, i.e., size, roof type, and age) was selected to collect sample of the study area. The selected building was easily accessible and was not close to an upwind or downwind source, to prevent interference with the typical natural process of atmospheric deposition of particulate matter in the area. The property is shown in Figure 4.2. It is a single storey building with a pitched, corrugated asbestos roof measuring 4 m × 14 m on each side of the pitch (Figure 4.3). The type of weather in the study site is typically tropical, with a succession of dry and rainy seasons.

The dry season usually runs from November to March while the rainy season generally starts in April, becomes fully established in May, and ends in October. The average yearly rainfall in the southern part of Nigeria (where the study area is located) varies from 1200 to 1500 mm (Adetunji et al., 2001; Adeniyi and Olabanji, 2005). Residential, commercial, retail, and public and private institutions are all part of the many land use activities that take place

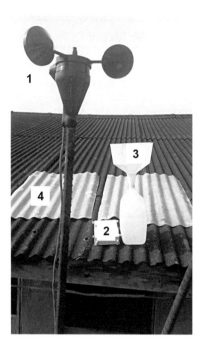

Figure 4.3 3-cup anemometer with roof-mounted modern device anemometer within 1 m – view 2. 1, 2, 3, and 4 denote the 3-cup anemometer, roof-mounted red hot anemometer, total deposition plastic container, and net deposition tiles, respectively

in Ikorodu's central regions. Over 43.5% of properties in the central region are used for commercial purposes, which constantly creates traffic and lively human activities (Bello, 2007a).

4.3 EXPERIMENTAL METHODS

4.3.1 Experimental Design

The designed experiment to analyse sampled rainwater was conducted between November and July of the following year, spanning both dry and rainy seasons[1]. For each sample, the methods outlined in Sections 3.2.4 and 3.2.5 were employed to continually monitor the wind speed and rainfall respectively. This allowed the recording of wind speed, dry antecedent days, and antecedent rainfall. The rain gauge was installed in a carefully selected site that was free from any obstacle to ensure reliable readings, and the chosen roof was clear of any adjacent impediments, including overhanging trees. The method described in Section 3.2.1 was also used to measure total deposition every week over the experiment's duration. The frequency

Table 4.1 Sampling plan for total deposited solids (g/m^2/week)

	Unfractionated samples	Fractionated samples
No. of sample points	1	*4
No. of months	8	8
Samples per month	3	1
Total no. of samples	24	32

Notes: *Each stage of the serial filtration was taken as a sampling point.

Table 4.2 Sampling plan for total coliforms & E. coli (MPN/100 mL) from total deposition

	Unfractionated samples	Fractionated samples
No. of sample points	1	4
No. of months	8	8
Samples per month	3	1
Duplicate samples	2	2
Set of dilutions	*1	*1
Total No. of samples	72	96

Notes: *Only one of the duplicate samples was diluted.

specified in Tables 4.1 and 4.2 was used to determine the deposit mass and bacteriological quality of the deposit. The collected sample was fractionated once a month using serial filtration. The distribution of deposit particle sizes and levels of bacterial contamination related to each size range were revealed by this filtration exercise.

4.3.2 Experiment in the Dry Season

November to February are typically dry months with little to no rainfall. Several series of roof tiles were exposed for different periods to determine how antecedent dry periods affected the accumulation of deposition. To ascertain if changes in the background environmental conditions during the experiment affected the accumulation outcome, the data on wind speed and total deposition were coherently analysed.

Four pre-sterilised 80 cm × 60 cm tiles were secured to the roof during the experiment. One tile was removed weekly, bi-weekly, and monthly for analysis on a rolling programme, and each tile was replaced immediately after removal. Therefore, a minimum of seven samples were collected per month (4 × weekly, 2 × bi-weekly, 1 × monthly, and 1 × bi-monthly) till the end of the dry season (refer to Tables 4.3 and 4.4). Tables 4.3 and 4.4 provide the frequency of sampling plan, while the monthly and weekly sample were

Table 4.3 Sampling plan for net deposited solids (g/m²/week) in dry period

	Unfractionated samples	Fractionated samples
No. of sample points	1	*4
No. of months	4	4
Samples per month	5	2
Total no. of samples	20	32

Notes: *Each stage of the serial filtration was taken as a sampling point.

Table 4.4 Sampling plan for total coliforms and *E. coli* (MPN/100 mL) from net deposition in the dry period

	Unfractionated samples	Fractionated samples
No. of sample points	1	4
No. of months	4	4
Samples per month	5	2
Duplicate samples	2	2
Set of dilutions	*1	*1
Total No. of samples	60	96

Notes: *Only one of the duplicate samples was diluted.

taken for fractionation testing. The tiles were washed with sterilised water after each analysis before being secured back on the roof to prevent external contamination that could corrupt the collected data.

The wind speed at the roof surface was also measured for a month using a hot-wire roof surface anemometer positioned on the roof (Figure 4.3). The results from both wind speed anemometer[2] and hot-wire roof surface anemometer were compared to quantify their measured result differences. This comparative analysis presents the degree of variations between the equipment. It is worth noting that it is not possible to operate the hot-wire anemometer in wet weather due to the unpredictable occurrence of rainfall events.

4.3.3 Experiment in the Rainy Season

Rainfall that occurs from April to July has a crucial impact on the accumulation of rooftop deposits. To study its impact, a gutter and storage tank were placed at the research property to collect and investigate the roof-harvested rainwater. For this purpose, aluminium box gutters were fabricated in the appropriate sizes to handle the anticipated high-intensity rainfall. The collection gutter was securely fastened to the fascia board with brackets, and a storage container was located at the end of the gutter's downpipe. Appendix B details the aluminium box gutter design.

Table 4.5 Sampling plan for roof-runoff solids (g/m²/week) in rainy period

	Unfractionated samples	Fractionated samples
No. of sample points	1	*4
No. of months	3	3
Samples per month	**8	2
Total no. of samples	24	24

Notes
*Each stage of the serial filtration was taken as a sampling point.
**The number of samples was dependent on the rainfall frequency; however, over 40 rainfall events were analysed.

The programme for measuring net deposition that was started during the dry season was continued throughout the whole rainy season. The goal of this lengthy measurement programme was to ascertain how much the roof's accumulation is removed by rain. In the proposed method, the tiles were fastened to the roof, and they were taken off immediately after a rainfall event[3], and was done throughout the rainy season[5]. Table 4.5 provides the experimental programme's frequency information.

A 255 L tank (standard and very common in the study area) was selected for measuring roof runoff deposit, being used to collect a full rainfall event of 50 mm/day from the designated roof area.[4] The analysis of each harvested rainfall event can be used to estimate the deposition per square metre as the gutter collects runoff from a defined roof area. The intensity of the rainfall was measured from the rain gauge following each rainfall event. The location of the rain gauge was free from any kind of obstruction, and the distance between the building roof and the rain gauge was kept to a maximum of 3 m to minimise external influence on the collected rainfall. A pre-sterilised sampling container was dipped into the centre of the storage vessel after being well mixed[5] to collect a 5 L water sample. Using the procedures outlined in Section 3.0, this 5 L water sample was sent to the laboratory for examination.

Table 4.6 Sampling plan for total coliforms & *E. coli* (MPN/100 mL) from roof runoff in rainy period

	Unfractionated samples	Fractionated samples
No. of sample points	1	4
No. of months	3	3
Samples per month	4	2
Duplicate samples	2	2
Set of dilutions	*1	*1
Total no. of samples	36	72

Notes: *Only one of the duplicate samples was diluted.

Table 4.7 Experimental results from the property in the dry season

Week	TD (g/m²/ week)	ND (g/m²/week)	% difference in the depositions	AWWS (MPH)	Period of exposure (days)	Rainfall (mm)
1	21.63	1.34	93.80	1.93	7	-
2	21.99	1.33	94.00	2.63	7	-
3	22.85	1.42	93.70	3.24	7	-
4	23.72	1.46	93.60	1.52	7	-
5	27.55	2.17	92.10	0.46	7	-
6	26.92	2.15	92.02	0.73	7	-
7	24.04	1.84	92.30	2.41	7	-
8	22.93	1.54	93.30	6.10	7	-
9	21.49	1.68	92.20	6.19	7	-
10	19.74	1.63	91.80	5.74	7	-
11	19.01	1.62	91.50	4.99	7	-
12	16.38	1.26	92.30	2.73	7	-
13	16.14	1.25	92.20	2.34	7	-
14	15.21	1.17	92.30	1.70	7	-

Note: TD, ND, %, and AWWS denotes total deposition, net deposition, percentage, and average weekly wind speed respectively.

A series of rainfall events was studied to determine how the rate of accumulation of deposition varies with an antecedent dry period, and data on wind speed and total deposition were continuously analysed to determine whether background environmental variables affected the experiment's results. Throughout the observed period, 44 rainfall events were studied (Table 4.8), which exceeds the number of samples required to calculate the unfractionated roof runoff solids given in Table 4.5. This part of the programme was extended till the end of the rainy period to allow firm conclusions due to the level of statistical significance between the outcomes.

4.3.4 Comparative Analysis of the Wind Speed from Both Recording Devices

Figure 4.4 describes the findings of the recorded wind speed from both the hot-wire anemometer and the Omega series data logging anemometer (main anemometer). Results showed that the wind speed (in MPH) for the hot-wire anemometer and the Omega anemometer varied from 0 to 35.89 and 0 to 46.16, respectively, while their respective average ranged from 0.01 to 7.12 and 0.012 to 9.91. The wind speed was recorded for 1 month (in January).

The average wind speed was variable for both recording devices, and the percentage difference ranged from 43.8% to 99.6%. The difference was

Table 4.8 Experimental results from the property in the rainy season

S/N	TRRND (g/m²/day)	Total deposition (g/m²/day)	DAD (days)	Rainfall intensity (inch/day)	Wind Speed (MPH)	% difference in the depositions
1	0.13	1.90	3	0.15	1.56	93.35
2	0.02	1.41	2	0.06	4.05	98.46
3	0.53	1.38	2	0.71	2.17	61.65
4	0.28	1.29	2	0.43	2.04	78.14
5	0.11	1.16	1	0.08	2.38	90.52
6	0.03	1.01	1	0.03	1.99	96.58
7	0.29	1.04	1	0.39	0.7	72.31
8	0.46	1.65	12	0.48	2.57	72.42
9	0.34	0.99	1	0.02	2.81	65.45
10	0.05	1.74	3	0.11	1.97	97.24
11	0.05	0.95	1	0.03	2.19	94.32
12	0.66	1.35	2	1.12	4.5	51.33
13	0.15	1.31	2	0.15	3.51	88.35
`14	0.13	1.22	1	0.09	2.5	89.67
15	0.45	1.66	8	0.42	2.26	72.75
16	0.22	1.40	1	0.3	2.56	84
17	0.34	1.45	4	0.32	0.82	76.72
18	0.03	1.25	2	0.05	1.59	97.75
19	0.03	1.59	9	0.13	4.19	98.25
20	0.4	1.26	1	0.43	1.99	68.45
21	0.01	1.13	2	0.07	3.77	98.74
22	0.74	1.20	2	1.44	3.34	37.76
23	0.27	0.93	1	0.72	1.21	71.45
24	0	0.61	1	0.37	1.44	100
25	0.14	0.67	1	0.18	2.57	78.51
26	0.09	0.72	2	0.16	2.86	86.88
27	0	0.51	1	0.23	2.72	100
28	0.44	0.63	1	0.87	2.72	30.87
29	0.77	1.62	5	1.03	3.74	52.56
30	0.02	0.47	1	0.1	5.61	95.91
31	0.44	0.99	5	1.03	5.89	56.09
32	0.55	0.68	2	0.49	1.54	19.15
33	0.54	0.89	6	0.76	6.28	39.36
34	0.12	0.45	1	0.15	5.51	73.11
35	0	0.49	1	0.26	6.91	100
36	0.28	0.57	2	0.64	5.76	51.58
37	0.02	0.41	1	0.07	2.19	96.1
38	0.02	0.38	1	0.02	1.23	94.2
39	0.29	0.43	1	0.06	0.73	31.63
40	0.02	1.35	10	0.12	4.27	98.24
41	0.18	0.39	1	1.03	2.24	53.33
42	0.02	1.04	4	0.06	8.96	98.44
43	0.05	0.90	5	0.02	5.83	93.89
44	0.2	0.51	2	0.55	7.99	61.29

Notes: S/N, TRRND, and DAD, depict serial number, total roof runoff net deposit and dry antecedent days respectively. The percentage difference in the depositions was obtained by dividing the differences in the depositions by the total deposition. Serial number denotes a rainfall event. TRRND is the total roof runoff net deposit in the rainy season.

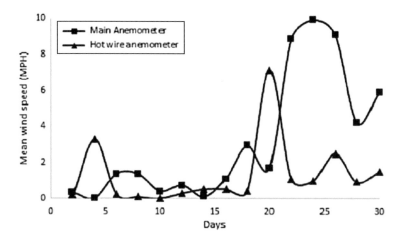

Figure 4.4 Comparative plots of the average speed for both wind speed recording
devices

caused by the various methods each piece of equipment used to record the
wind speed. The main anemometer had a lower sampling frequency and
measured wind speed in minutes, whilst the hot-wire anemometer measured
wind speed in seconds. On the other hand, there is a tendency for the data
from the hot-wire anemometer to be either too high or too low since more
data is captured. The hot-wire anemometer operates with a 5-volt battery;
therefore, there is a chance that the battery may record inaccurate/incom-
plete data, particularly when the battery's voltage is running low after being
exhausted by high sampling demand.

In comparison, the Omega series data logging anemometer can be used
for both short and long-term wind speed data recording; however, the hot-
wire anemometer can only be used for short-term wind speed recording
since it relies on rechargeable AA batteries (due to the unavailability of
other safe and reliable power sources). Also, the results show that hot-wire
anemometers are operationally unreliable due to their high fragility and
the rapid rate at which their calibration varies in unclean or wet/damp
environments.

4.4 RESULTS AND ANALYSIS

4.4.1 Deposition From a Property

In this section, results from the samples and experimental analysis – including
the weekly experimental deposition, rainfall intensity, wind speed, impact of
the dry antecedent days, and depositions in both seasons – are presented. The

accumulation of the deposits in the short term (3 and 7 days) and long term (14, 28, and 56 days) are also analysed and compared.

Deposition from the Investigated Property in the Dry Season

Table 4.7 presents the findings during the dry season, i.e., total deposition, net deposition, and weekly average wind speed measured during the 14-week experimental period. The commencement of the rainy season prevented the weekly total and net deposition from continuing after 14 weeks. The recording of wind speed continued after the 14 weeks to enable the multiple linear regression of the involved variables (i.e., total deposition, roof runoff deposit, wind speed, and amount of rainfall as recorded by rain gauge). The total and net deposition of particulate matter (g/m²/week) varied from 15.21 to 27.55 and from 1.17 to 2.17, respectively, throughout the dry season, with their corresponding averages of 21.4 and 1.56 (refer to Table 4.7). Weeks 5 and 14 had the largest and lowest total and net depositions in the dry season, respectively (Table 4.7 and Figure 4.5). The peak deposition rate was found during the Harmattan period, whilst the least amount was observed at the end of the dry season. Harmattan periods usually cause desert-like weather conditions. Occasionally, dust storms and sandstorms can be formed to severely obstruct visibility (Sunnu et al., 2013). Furthermore, they reduce humidity, disperse cloud cover, and prevent rain clouds from developing. The percentage differences between the total and net depositions ranged from 91.50% to 94% (Table 4.7).

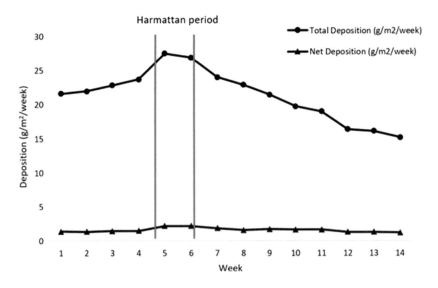

Figure 4.5 Plot of the depositions from the property in the dry season

The significant difference between total and net deposition is caused by the influence of wind. The range of weekly average wind speed (MPH) in the dry season was 0.46 to 6.19, with an average of 3.05, whereas the least wind speed was recorded during the Harmattan period (Table 4.7). Net and total depositions continuously reduced after the Harmattan period (Figure 4.5), and this occurred because of the weather shift towards the rainy season, which suggests that there were less suspended materials in the atmosphere at that period.

Deposition from the Investigated Property in the Rainy Season

The range of the average weekly total and net depositions (g/m²/day) in the rainy season were 0.38–1.9 and 0–0.77, respectively; while their respective averages were 1.02 and 0.23 g/m²/day (refer to Table 4.10 and Appendix C). Unlike the dry season when the depositions were analysed every week, the rainy season's measurements were dependent on when the rainfall event took place. Appendix C gives comprehensive results of the involved parameters during the rainy season. The period of dry antecedent days recorded before the rainy season was between 1 and 12 days. The amount of rain recorded during the measured period ranged from 0 mm to 197.87 mm (total of 644.66 mm), while the lowest and highest amounts of rain were recorded during December and June, respectively (Figure 4.6).

The total deposition of particulate matter was higher in the dry season than the rainy season (Table 4.9). It decreased as the rainy season progressed

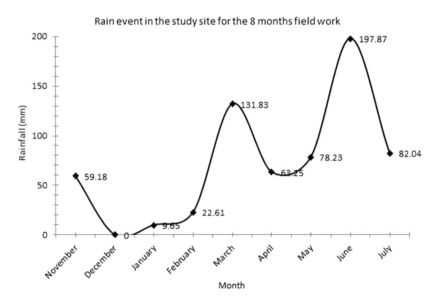

Figure 4.6 Amount of rain recorded for the 8 months of field data

Table 4.9 Statistical analysis of total deposition
(g/m²/week) results in both seasons

	Minimum	Maximum	Average
DS	15.21	27.55	21.4
RS	2.66	13.3	7.14

Note: DS and RS denote dry season and rainy season
respectively.

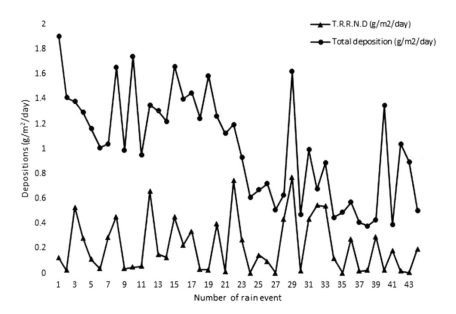

Figure 4.7 Total deposition and Total Roof Run-off Net Deposit (TRRND) per day in the rainy season.

(Figure 4.8). This phenomenon was also suggested by some studies, as the rain reduced the accumulated mass of particulate matter in the atmosphere (Moberg et al., 1991; Adetunji et al., 2001).

4.4.2 Statistical Analysis of the Depositions in Both Seasons

This section employs the Multiple Linear Regressions (MLR) model to analyse depositions during the rainy and dry seasons. Using measured data to best fit an equation, the MLR can describe the relationship between two or more explanatory factors and a response variable.

Statistical Analysis of Results from the Property in the Dry Season

The explanatory factors in the rainy seasons are dry antecedent days (DAD), wind speed, total deposition, and rainfall intensity, whereas these variables in the dry season are average weekly wind speed (AWWS), total deposition, and DAD. The representative equations (from all explanatory factors) can illustrate the interaction between the various independent factors and the dependent variable (i.e., net deposition). Tables 4.10a and 4.10b present the results of the MLR analysis during the dry season, and Table 4.9 describes the data used for the MLR analysis in the dry season. The obtained equation from the summary output of Table 4.10a is shown below:

$$\text{ND (g/m}^2\text{/week)} = 0.07 + 0.068 \text{ TD (g/m}^2\text{/week)} + 0.01 \text{ AWWS (MPH)} \quad (4.1)$$

To investigate deposition in the dry season, the R-square and the adjusted R-square have been used, where R-square is a measure that indicates how well a model fits the data. It is a statistical indicator of how closely the regression line resembles the actual data in the context of regression, while the adjusted R-square is a variant of R-square that considers factors in a regression model that are not significant. To put it another way, the adjusted R-square demonstrates whether the regression model is made better by including or removing more factors. The R-square and the adjusted R-square were obtained as part of the output from the MLR analysis, and their values are 64.7% and 58.2%, respectively (Table 4.10a). Due to the non-linear interaction between the independent variables by the adjusted R-square, it is more reliable in the MLR analysis. The MLR analysis suggests that total deposition and AWWS (i.e., the two independent variables) account for 58.2% of the variability in the dependant variable (i.e., net deposition). The p-value in MLR analysis is used to describe whether there is a statistically significant relationship between each predictor variable and the response variable. The null hypothesis is disregarded if the p-value is less than 0.05 since there are significant differences in the variable means, whereas significant differences do not exist if the p-value is more than 0.05. Further analysis indicates that only the p-value for total deposition was less than 0.05, according to the Analysis of Variation (ANOVA) for the data in Table 4.10a. This implies that only the coefficient values (in Table 4.10b) for total deposition are statistically significant. The MLR was once more performed utilising only the total deposition and net deposition data from Table 4.8, and the summary output is presented in Table 4.10b.

An adjusted R-square of 61.3% was obtained from the MLR results utilising the data on net and total depositions (and not the AWWS data). Additionally, it was noted that the p-value for total deposition is less than 0.05; while the p-value for the intercept (i.e., 0.688 in Table 10b) is larger than 0.05. This finding suggests that the MLR equation applies relatively

Table 4.10a Summary output of the multiple linear regression (MLR) analysis in the dry season

Regression Statistics

Multiple R	0.804
R Square	0.647
Adjusted R Square	0.582
Standard Error	0.204
Observations	14

ANOVA

	Df	SS	MS	F	Significance F
Regression	2	0.835	0.417	10.061	0.003
Residual	11	0.456	0.041		
Total	13	1.291			

	Coefficients	Standard Error	t Stat	P-value	Lower 95%	Upper 95%
Intercept	0.070	0.370	0.188	0.855	-0.746	0.885
TD (g/m2/week)	0.068	0.015	4.407	0.001	0.034	0.102
AWWS (MPH)	0.010	0.030	0.331	0.747	-0.056	0.076

Table 4.10b Summary output of the multiple linear regression analysis in the dry season

Regression Statistics

Multiple R	0.802
R-square	0.643
Adjusted R-square	0.613
Standard Error	0.196
Observations	14

ANOVA

	Df	SS	MS	F	Significance F
Regression	2	0.835	0.417	10.061	0.003
Residual	11	0.456	0.041		
Total	13	1.291			

	Coefficients	Standard Error	t Stat	P-value	Lower 95%	Upper 95%
Intercept	0.128	0.313	0.411	0.688	-0.553	0.810
TD (g/m²/week)	0.067	0.014	4.649	0.001	0.036	0.078

well to the coefficient of total deposition. After the second MLR was done (Table 4.10b), the formulated equation becomes

$$\text{ND (g/m}^2\text{/week)} = 0.067 \text{ TD (g/m}^2\text{/week)} \tag{4.2}$$

It can be concluded from Equation 4.2 that the major variable factor that influences the net deposition in the dry season is the total deposition, because the regression shows that there is a corresponding 0.067-unit gain in net deposition for every unit increase in total deposition during the dry season.

Statistical Analysis of Results from the Investigated Property in the Rainy Season

The data utilised and result summary for this MLR analysis are shown in Tables 4.8 and 4.11 respectively. The dependent variables employed include DAD, wind speed (ADWS), total deposition, and rainfall intensity, while the independent variable is the total roof runoff net deposit (TRRND) in the rainy season. TRRND determines the volume of particulate matter removed from a gutter catchment area in each rainfall event. The full data are included in Appendix C, which contains a breakdown of all the measurement findings from the property throughout the rainy season as well as the step-by-step calculations used to derive all necessary variables for Table 4.8.

The regression equation obtained from the MLR is:

$$\text{TRRND (g/m}^2\text{/day)} = -0.009 + 0.089 \text{ TD (g/m}^2\text{/day)} + 0.05 \text{ DAD}$$
$$\text{(days)} + 0.483 \text{ RI (inch/day)} - 0.014 \text{ ADWS (MPH)} \tag{4.3}$$

Table 4.11a shows that the R-square and modified R-square for this analysis are 70.9% and 67.9%, respectively, which demonstrates that the four independent variables of total deposition, DAD, rainfall intensity, and ADWS account for 67.9% of the fluctuations in the dependent variable (i.e., net deposition). The ANOVA analysis outcome in Table 4.11a reveals that only the coefficient values of rainfall intensity are statistically significant since their p-value is less than 0.05. Therefore, the MLR was once more performed utilising only the information from the rainfall intensity in Table 4.8. The findings of the MLR utilising only the data from the TRRND and rainfall intensity in Table 4.10 present 63.3% of the variation in the TRRND variables. It suggests that only the coefficient of rainfall intensity is valid for the MLR equation.

After the second MLR was done (Table 4.13b), the best-fit equation becomes

$$\text{TRRND (g/m}^2\text{/day)} = 0.483 \text{ RI (inch/day)} \tag{4.4}$$

Table 4.11a Summary output of the multiple linear regression (MLR) analysis in the rainy season

Regression statistics

Multiple R	0.842
R-square	0.709
Adjusted R-square	0.679
Standard Error	0.124
Observations	44

ANOVA

	Df	SS	MS	F	Significance F
Regression	4	1.455	0.364	23.722	5.31E-10
Residual	39	0.598	0.015		
Total	43	2.053			

	Coefficients	Standard Error	t Stat	P-value	Lower 95%	Upper 95%
Intercept	-0.009	0.068	-0.13	0.897	-0.147	0.129
TD (g/m²/week)	0.089	0.055	1.62	0.113	-0.022	0.201
DAD (days)	0.005	0.009	0.582	0.564	-0.013	0.023
Rainfall intensity (inch/day)	0.483	0.052	9.206	2.00E-11	0.377	0.589
ADWS (MPH)	-0.014	0.011	-1.343	0.187	-0.035	0.007

Table 4.11b Summary output of the multiple linear regression analysis in the rainy season

Regression Statistics

Multiple R	0.801
R-square	0.642
Adjusted R-square	0.633
Standard Error	0.132
Observations	44

ANOVA

	Df	SS	MS	F	Significance F
Regression	4	1.318	1.318	75.3	6.48E-11
Residual	39	0.735	0.018		
Total	43	2.053			

	Coefficients	Standard Error	t Stat	p-value	Lower 95%	Upper 95%
Intercept	0.05	0.028	1.776	0.083	-0.007	0.108
Rainfall intensity (inch/day)	0.483	0.056	8.678	6.00E-11	0.371	0.595

According to Equation 4.4, the total roof runoff net deposition increases by 0.483 units for every unit increase in rainfall intensity throughout the rainy season. The main variable factor that affects the total roof runoff net deposition during the rainy season is the intensity of rainfall (Equation 4.4).

4.4.3 Serial Filtration of the Depositions

Total and net depositions were fractionated in both seasons in this study. The findings of the fractionation tests, which were used to define the range of each particle size, are shown and discussed in this section.

Serial Filtration of the Depositions and its Associated Bacteria in the Dry Season

According to the fractionation of the total deposits made during the dry season, the proportion of solids deposited in 10 μm and 500 μm filters ranged from 69% to 75.1% and 5.2% to 9.2%, respectively, their respective averages being 72.1% and 7.3% (Table 4.12). Additionally, the fractionation of the net deposits during the dry season reveals that the averages and ranges of the solid percentages deposited in the 10 μm and 500 μm filters were 6.0 and 74.3, as well as 4.7 to 8.1 and 69.8 to 77.2 (refer to Table 4.13). The results also showed that not all the solids in the raw samples were retained after serial filtration due to some solids having a size smaller than a 10 μm filter.

Due to the filter's pore size, the 10 μm filter effectively retained most of the deposits, whereas the 500 μm filter retained the least amount of materials (Tables 4.12 and 4.13). This was consistent for all the performed fractionation experiments. In Appendix D, Tables D1, D2, and D3 show the fractionation results for both seasons. Table D1 presents the fractionation results of total and net depositions and their associated bacteria during the dry season, while Tables D2 and D3 illustrate those results during the rainy season. The

Table 4.12 Percentage of solids (from total deposition) retained by the filters in the dry season

S/N	Percentage retained by 500 μm filter	Percentage retained by 100 μm filter	Percentage retained by 10 μm filter	Percentage of solids < 10 μm filter
1	8.7	10.5	72	8.8
2	9	11.4	72.1	7.5
3	9.2	12.6	69	9.2
4	6.9	11	69.1	13
5	6.2	9.7	74.1	10
6	6.1	8.5	75.1	10.3
7	5.2	7.8	73.1	13.9
Average	7.3	10.2	72.1	10.4

Table 4.13 Percentage of solids (from net deposition) retained by the filters in the dry season

S/N	Percentage retained by 500 µm filter	Percentage retained by 100 µm filter	Percentage retained by 10 µm filter	Percentage of solids < 10 µm filter
1	4.7	9.6	76.6	9.1
2	5.8	9.8	74.5	9.9
3	8.1	12.8	69.8	9.3
4	7	11.3	70.9	10.8
5	6.2	10.5	77.2	6.1
6	5.5	9.9	74.6	10
7	5	9.6	76.4	9
Average	6	10.5	74.3	9.2

finding from the fractionation experiments in Appendix D showed similar patterns for the duplicates and diluted tests.

The microbiological examination of the serial filtrate shows that little to no bacteria (from 0% to 2%) were eliminated after it was passed through the mesh of the 500 µm filter. In comparison, more bacteria (4%–7%) were removed after it was passed through the mesh of the 100 µm filter. The 10 µm filter paper removes a significant amount of bacteria (between 10% and 45%) when compared to the 500 µm and 100 µm filter papers. Since some bacteria are attached to particles, they can be eliminated when the solids are removed in the filter during the serial filtration process. However, in general, bacteria removal using mesh and filters can be classified as ineffective. The mass of solids deposited in the fourth and fifth fractionation experiments was lower than those in the first three. The majority (about 70%) of samples from the examined deposits fall between the sizes of 10 µm and 100 µm. The peak of the Harmattan period can be connected to the rise in solids, and the subsequent continuous fall strengthens the conclusion for the dry season.

Serial Filtration of the Depositions in the Rainy Season

The fractionation of the rainy season samples revealed that the percentage of solids (from the total deposition experiment) retained by the 100 µm filter ranged from 5.5 to 7.7 with an average of 6.5. In contrast, the percentage of solids retained by the 500 µm mesh filter ranged from 2.7 to 4.6 with an average of 3.6. Additionally, the fractionation test findings revealed that the percentage of solids deposited in the 10 m filter ranged from 67.6 to 71.3 with an average of 69.8, while some extremely small particles were unaccounted for. With an average of 20.1%, the percentage of these unmeasurable solids ranged from 16.4 to 24.2 (Table 4.14).

Table 4.14 Percentage of solids (from total deposition) retained by the filters in the rainy season

S/N	Percentage retained by 500 μm filter	Percentage retained by 100 μm filter	Percentage retained by 10 μm filter	Percentage of solids < 10 μm filter
1	4.6	7.7	71.3	16.4
2	4	7	70.7	18.3
3	3.8	6.4	69.1	20.7
4	2.7	6.1	70.4	20.8
5	2.7	5.5	67.6	24.2
Average	3.6	6.5	69.8	20.1

On the other hand, during the dry season, the results of the fractionation of the samples revealed that the percentage of solids from the roof runoff net deposit retained by the 100 μm filter ranged from 11.3 to 28.6 with an average of 17.7, while that retained by the 500 μm mesh filter ranged from 1.7 to 8.7 with an average of 5.7. The findings from the sample fractionation showed that the percentage of the solids (from the roof runoff net deposit) deposited in the 10 μm filter during the rainy season ranged from 44.9 to 75.8 with an average of 69.8. The percentage of these unaccounted solids ranged from 1 to 30 with an average of 12, which is less than the rainy season (Table 4.15).

Table 4.15 Percentage of solids (from roof runoff deposit) retained by the filters in the rainy season

S/N	Percentage retained by 500μm filter	Percentage retained by 100μm filter	Percentage retained by 10μm filter	Percentage of solids < 10 μm filter
1	3.6	13	71.7	11.7
2	1.7	23.7	47.5	27.1
3	6.4	13	67.8	12.8
4	7	11.3	75.8	5.9
5	6.2	15.2	71	7.6
6	7.3	13.3	68.8	10.6
7	8.7	27.5	44.9	18.9
8	5.7	28.6	60	5.7
9	4.2	16.4	72	7.4
10	5.7	17.5	71.6	5.2
11	7.9	19.8	71.3	1
12	4.2	12.9	52.9	30
Average	5.7	17.7	64.6	12

Reflecting on the results from Tables 4.14 and 4.15, it can be seen that a 10 µm filter retained most of the samples throughout the rainy season, while the 500 µm filter sieved the least amount. This was consistent for all the fractionation experiments that were performed in both rainy and dry seasons. As evident from the previous sections, there are more depositions in the dry season than in the rainy season, and this is due to longer undisturbed periods of deposition of particulate matter from the atmosphere.

4.4.4 Accumulation of Deposition over Time

The measured net depositions were correlated with the different exposure days to determine the significance of deposit accumulation over time. This correlation is imperative as it can be associated with the attached bacteria to particulate matter. These depositions consist of both long-term and short-term net depositions. The measurement of the 7-day (weekly) and the 3-day net deposits are the short-term net deposits, whereas the measurement of the fortnightly (14 days), monthly (28 days), and bi-monthly (56 days) net deposits are the long-term net deposits.[6] Results for both the long and short-term net depositions that took place throughout the dry season are shown in Table 4.16. Most short-term depositions were sampled during the rainy season since rainfall events washed away the depositions. The correlation analysis revealed a strong positive correlation of 0.92 between the accumulated net depositions and the period of exposure (Table 4.16 and Figure 4.8). The linear equation gives

$$\text{Accumulated Net Deposit (g/m}^2\text{/day)} = 0.151 \text{ DAD (days)} + 0.1023 \quad (4.5)$$

According to Equation 4.5, there will be a 0.151-unit increase in the accumulated net deposit for every 1-unit increase in the number of dry antecedent days. The finding indicates that there were higher accumulated deposits during the dry season than during the rainy season (refer to Table 4.16), which can be explained by the prolonged dry antecedent period in the dry season. S/N in Table 4.16 denotes a serial number, which is a number assigned to each deposition experiment, from 1 to 64. The results from 1 to 40 were completed in the dry season, and the remaining ones (i.e., from 41 to 64) were completed in the rainy season (Table 4.16).

4.5 DISCUSSION

The impact of particulate matter deposition on the quality of harvested rainwater has been investigated, and results show that particles exist in the atmosphere and are deposited on the building roofs. It was also evidenced that some microorganisms are attached to these particles, therefore making

Table 4.16 Accumulated net deposits over both seasons

S/N	Period of exposure (days)	Accumulated deposit (g/m²)
1	14	2.49
2	14	2.55
3	14	3.18
4	14	2.72
5	28	3.63
6	28	4.51
7	56	7.88
8	3	0.58
9	3	0.6
10	3	0.62
11	3	0.53
12	7	1.34
13	7	1.33
14	7	1.42
15	7	1.46
16	7	2.17
17	7	2.15
18	7	1.84
19	7	1.54
20	3	0.53
21	3	0.55
22	3	0.5
23	7	1.68
24	3	0.46
25	3	0.47
26	3	0.49
27	3	0.44
28	3	0.42
29	7	1.63
30	7	1.62
31	14	2.58
32	3	0.41
33	3	0.43
34	3	0.43
35	3	0.46
36	7	1.26
37	3	0.39
38	3	0.41
39	3	0.43
40	3	0.42
41	7	1.25
42	3	0.35
43	3	0.38
44	3	0.38
45	7	1.17
46	3	0.35

(Continued)

Table 4.16 (Continued)

S/N	Period of exposure (days)	Accumulated deposit (g/m²)
47	3	0.34
48	3	0.34
49	3	0.35
50	7	1.15
51	3	0.3
52	3	0.31
53	3	0.29
54	3	0.28
55	3	0.27
56	3	0.25
57	3	0.26
58	3	0.27
59	7	1.13
60	3	0.22
61	3	0.24
62	3	0.21
63	3	0.22
64	3	0.22

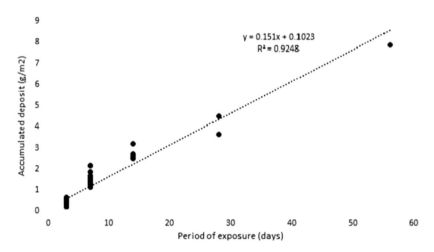

Figure 4.8 Correlation between the total solids (g) and enumerated total coliform (MPN/100 mL) for net deposition

building roofs a critical pathway for sediment mixing with rainwater. In both dry and rainy seasons' atmospheric deposition, microorganisms and trace metals bonded to various sized particulate matter are influenced by anthropogenic and natural sources, which over time may have significant impacts on ecosystems and human health. There are fewer studies that summarise the origins, routes, and global concentrations of atmospheric particulate matter depositions (Grantz et al., 2003; John et al., 2021a). The concentration of bacteria and fungi attached to the particulate matter have been recorded in different seasons by several studies in different developing and under-developed countries (Characklis et al., 2005; Krometis et al., 2013; Morakinyo et al., 2019; John et al., 2021c, Vithanage et al., 2022), and these contaminated particles may have adverse pollution effect when they enter the storage tanks. Conclusively, those studies found that the number of microorganisms in the harvested rainwater was found to exceed the WHO guidelines for drinking water.

Deposition on surfaces (like building roofs) promotes the accumulation of solids, metals, and organic material, which is transported to rainwater storage tanks after a rainfall event. The results from the measurements provided by this study and other research have demonstrated that atmospheric deposition is one of the primary sources of contamination in roof-harvested rainwater (Grantz et al., 2003; Prospero and Arimoto, 2009; Revitt et al., 2014; John et al., 2021c). The quality of rainwater is significantly impacted by the sources of particulate matter. From degraded leaf detritus to plant waxes, plants create a vast variety of organic particles. Particulate matter is also heavily enriched with trace metal pollutants that are released by smelters, power plants, and incinerators. Numerous organic compounds are released into the air by sources of combustion, including both anthropogenic and natural sources, including power plants, cars, biomass burning, and others. Particles rich in nitrogen are produced because of fertiliser use and combustion processes. From household and commercial sources, pesticides and other synthetic organic chemicals are released (Prospero and Arimoto, 2009). Traffic, atmospheric deposition, roofing materials, street furniture (such as lighting, barriers, and signage), general litter, and erosion of soil from nearby areas via both direct and indirect deposition routes are additional sources of particulate matter that can end up on building roofs (Gunawardana et al., 2012; Revitt et al., 2014).

Several variables, including the kind of catchment surface, the intensity and frequency of rainfall event, and the antecedent conditions, can greatly affect the quality of rainfall runoff. The cleanliness and hygiene of the regions used for collecting rainwater, including the gutter, pipelines, and closeness to animal waste, are other important aspects that influence its quality (John et al., 2021). The intensity and frequency of rainfall events are frequently included in generating models to characterise the accumulation of solids and their corresponding contaminants on urban surfaces. It is regarded by many researchers as a significant variable. The following wash-off of surface deposited pollutants makes a considerable contribution to diffuse pollution loads, which must be reduced in accordance with the challenging goals set forth by the EU Water Framework Directive (EU WFD, 2000; Revitt et al., 2014).

Rainfall characteristics influence deposition and pollution levels (Revitt et al., 2014). As shown in the results in this study, the rainfall intensity had the most significant impact on deposition, especially in the rainy season, and this agrees with Revitt's finding.

The atmospheric deposition of particulate matter has been linked to acid rain and climate change. Particulate matter deposition can exacerbate acid rain, especially in areas where there is gaseous emission. It also alters weather patterns, causes drought, accelerates global climate change, and causes the ocean to become more acidic. The impact of particulate matter deposition is evident throughout the sub-Saharan region of Africa and the Northern Hemisphere. In the Northern Hemisphere, high concentrations of SO_4^{2-} and NO_3^- are observed in the western Pacific near Asia, which is caused by continental outflow and made worse by the lack of effective emission regulations. As a result of emissions from North America and Europe, moderately high pollution levels are also observed over the North Atlantic. On the other hand, the concentrations of SO_4^{2-} and NO_3^- are incredibly low in American Samoa and the Antarctic stations of Mawson and Palmer. These conditions are what one should anticipate when pollution impacts are the least (Prospero and Arimoto, 2009; Revitt et al., 2014).

In the sub-Saharan regions of Africa, it is known that during the dry season (which lasts from November to March every year[7]), Saharan particulate matter occasionally enters the West African nations near to the Bay of Guinea. The Harmattan, a peak period of particulate matter deposition in the sub-Saharan region of Africa, typically occurs between December and January each year, is a key period of particle transport and settlement on roofs. The Bodélé Depression in the Chad Basin, which is supposedly the lowest place in Chad, is located at the southern tip of the Sahara Desert in the country's north central area. It is the source of the dry and hazy dust (Grantz et al., 2003; Prospero, 2011; Sunnu et al., 2013; John et al., 2021; Vithanage et al., 2022). These studies elucidated the impact of atmospheric deposition with respect to the transportation of gases and its impact on weather pattern alteration.

The wind speed is widely varied, ranging from 0.46 to 6.19 MPH, as shown in Table 4.7. According to Alghamdi et al. (2014), wind speed has a positive correlation with bioaerosol-bound particulate matter but has a negative correlation with the magnitude of particulate matter (Georgakopoulos et al., 2009). Bioaerosol-bound particulate matter is made up of extremely tiny airborne particles (from 0.001 to 100 μm) that are biologically derived from plants or animals and may contain living creatures.[8] The outcome from this study's measurements agrees with Alghamdi et al. (2014) and Morakinyo et al. (2019), which shows that wind speed is positively correlated with bioaerosol-bound particulate matter. The survival of airborne microorganisms depends on the wind speed. The bioaerosol-bound particulate matter is transported from the source to the sample location, and this lowers their net concentration by acting as a dilution factor through diffusion.

It can be concluded that building roofs represent a crucial pathway for particulate matter to enter rainwater storage tanks. Since microorganisms can be attached to particulate matter, there is the possibility that they will enter the rainwater storage tank. It is crucial to ensure that harvested rainwater is disinfected since microorganisms like bacteria and fungi might be attached to the deposited particulate matters (Morakinyo et al., 2019; John et al., 2021b). The severity of unfavourable health effects in humans depends heavily on the degree of the absorbed dose of pathogens (Morakinyo et al., 2019). Compared to adults, children less than 6 years old are more vulnerable to the negative impacts of airborne deposition because of their immature or under-developed lungs and their increased physical activity and greater breathing rates (Morakinyo et al., 2019).

4.6 CONCLUSIONS

From the results obtained in the measurements performed in this chapter, the primary conclusions of this chapter are as follows:

- Compared to the rainy season, the dry season has a higher total deposition. The significantly higher concentration of particles in the air during the dry season is a result of the longer dry antecedent period and reduced cloudiness in the atmosphere. The findings also indicated that the Harmattan period in the dry season had the highest deposition, which could be explained by the desert-like meteorological conditions.
- In both the rainy and dry seasons, the analysis of Multiple Linear Regressions revealed that rainfall intensity and total deposition are the main factors influencing the quantity of particulate matters. After using information from the total deposition and wind speed for the MLR study, the regression analysis found that the total deposition had an impact on net deposition during the dry season. However, during the rainy season, the intensity of the rainfall had an impact on net deposition.
- The fractionation of the sampled deposits revealed that majority of the samples (about 70%) fall between 10 µm and 100 µm in size, while the mesh size of 500 µm recorded the fewest deposits. However, some of the solids were unaccounted for, and this is because they were smaller than the 10 µm filter pore size.
- Most of the removed bacteria during the serial filtration processes were caught by the 10 µm filter paper. The removed bacteria from the serial filtration process proves that some bacteria are attached to solids, so as the solids were removed by the filter and meshes, the bacteria in the sample reduced.
- The findings from the correlation of deposits indicated that there is an accumulation of deposits over time (especially in the dry season). This suggested that most rainfall events that occur after long periods of dry

antecedent days are likely to have a higher record of deposits (leading to higher bacteria counts). The results further depicted that there was more accumulation of deposits in the dry season than in the rainy season.

NOTES

1 The dry season is defined as the November–December–January–February period, while the rainy season is defined as the April–May–June–July period based on the local Ikorodu weather.
2 The wind speed anemometer was positioned atop a 3 m pole close to the property.
3 It was anticipated that the tiles will be secured to the roof and will experience an antecedent dry period of at least 5 days prior to the rainfall event (in the dry season). The exact nature of the experimental design was dependent on the results of the work completed during the dry period.
4 The gutter collected water from 0.5 m of roof line, i.e., over an area of 0.5 m × 4 m. Assuming the 50 mm/day is a significant event, then a 255 L tank gives a factor of safety of 2. This was appropriate considering the uncertainty around rainfall data.
5 In the event of a very small rainfall event (i.e., less than 2 L), the entire volume of collected rainfall was filtered.
6 The measurement of the 7-day (weekly) and the 3-day net deposits are the short-term net deposits, whereas the measurement of the fortnightly (14 days), monthly (28 days), and bimonthly (56 days) net deposits are the long-term net deposits.
7 The dry season lasts from November to March every year.
8 Bioaerosol-bound particulate matter is made up of extremely tiny airborne particles (from 0.001 to 100 μm) that are biologically derived from plants or animals and may contain living creatures.

Chapter 5

Sedimentation and Rainwater Quality in Storage Tanks

ABSTRACT

Harvested rainwater has the potential of being an improved source of drinking water. The challenge observed with rainwater is the possible contamination which can emanate from sources due to atmospheric deposition, hygiene, and sanitation. As a result, it is imperative to investigate ways and implement measures to improve the quality of harvested rainwater. Using a case study of the Ikorodu area of Lagos state, Nigeria, this chapter examines the effect of sedimentation on rainwater storage systems. The proportions of Escherichia coli (*E. coli*) that settle (owing to sedimentation) and those that remain in the free phase are also determined. For a period of 20 days, water samples were taken from various depths in the investigated rainwater storage tank in both rainy and dry periods. In addition, the samples were tested through a series of filters with pore diameters of 500 µm, 100 µm, 10 µm, and 1.5 µm to quantify the retained particle mass, which was then examined for physical and microbiological parameters.[1] The outcomes demonstrated that the settle-able bacteria quickly sank to the bottom of the tank leading to improved water quality in the free phase zone. The physical sedimentation process was shown to significantly reduce the microbiological parameters since over 70% of the settling total suspended solids (TSS) occurred within the first 36 hours of settlement.

5.1 INTRODUCTION

Shortage of safe drinking water is one of the numerous significant issues that are being experienced in the world today. As the world's population and water demand have increased within the last half-millennium, the literature shows that water scarcity can be a major challenge (Matondo et al., 2005; Kaldellis and Kondili, 2007; Lee et al., 2010; DFID, 2015; WaterAid, 2015; WHO/ UNICEF JMP, 2015). The utilisation of harvested rainwater has the propensity to alleviate the continuing shortage of water, especially in less economically developed countries. Presently, several studies in the literature have considered harvested rainwater as a solution to drinkable water issues (Hatibu et al., 2006; Hartung, 2007; Ghisi and Ferreira, 2007; Lee et al., 2010 DFID, 2015;

DOI: 10.1201/9781003392576-5

WaterAid, 2015; WHO/UNICEF JMP, 2015; John et al., 2021c,d). However, the exploitation of harvested rainwater has been underutilised in most developing countries as mostly only individual households harvest rainwater.

One of the major issues with respect to the harvesting and use of roof-harvested rainwater is the presence of contaminants (e.g., physicochemical contaminants or microbiological pathogens). Atmospheric influences on harvested rainwater and the impact of the age and cleanliness of storage tanks, gutters, roof catchments, and collection pipes are two major sources of external contamination identified by various researchers as affecting the quality of harvested rainwater (Yaziz et al., 1989; Simmons et al., 2001; Chang et al., 2004; Zhu et al., 2004; Sazakli et al., 2007; Lee et al., 2010). This chapter will investigate the level to which bacterial indicator organisms in water storage tanks are associated with settle-able particles. Since the study conducted by John et al. (2021c) on the study area showed that the residents harvest and store rainwater for drinking, it is imperative to investigate the fate of the microbes in the storage tank, focusing on specific mechanisms of water quality improvement such as sedimentation.

One of the processes used in the treatment of water is sedimentation, which employs gravity to remove suspended materials from water. Coagulation and filtration are the phases before and after the sedimentation stage in the water treatment process. Sedimentation occurs naturally in the still water of lakes and seas to settle solid particles that are entrained by the turbulence of moving water. Ponds called settling basins are made with the intention of using sedimentation to remove entrained sediments. Tanks designed with mechanical mechanisms for the continual removal of solids being deposited by sedimentation are called *clarifiers*, while *sludge* is the term for the sedimentary particles that form from suspension during the water treatment process (Omelia, 1998). The consolidation process occurs when there is a continuous settlement of a heavy layer of sediment, while thickening is a process of assisting sediment or sludge in consolidating by mechanical means (Gregory and Edzwald, 2010).

The sedimentation process occurs in non-flowing water, like a rainwater storage tank, because gravity drags heavier sediments down to produce a sludge layer. Studies have revealed that the addition of coagulants, such as aluminium sulphate (alum), tend to neutralise the negatively charged particles and weaken the forces holding colloids apart and can speed up the sedimentation process (Zhou et al., 2012; Hussain et al., 2014; John et al., 2021b). According to Characklis et al. (2005), Krometis et al. (2013), and John et al. (2021b), microbes such as viruses, bacteria, and protozoa can attach to suspended solids that have more unpredictable transport characteristics compared to bed-load solids (Pu, 2021; Pu et al., 2014; and Pu and Lim, 2014). Higher turbidity levels can also prevent microorganisms from the effects of chlorination and promote bacterial development, which increases the amount of chlorine needed in water treatment (Sharma & Bhattacharya, 2017). Since they are free floating in the water, the free phase bacteria in a water storage tank can be eliminated by boiling the water or applying disinfectants, while the sessile bacteria can be significantly reduced by the sedimentation process

(Olson et al., 2002; Characklis et al., 2005; Krometis et al., 2013; John et al., 2021c). To perform the necessary knowledge connections between the sedimentation process and water quality, the level of settle-able particles in the rainwater storage tank will be examined in this chapter along with its correlation with the measured parameters (i.e., physicochemical and microbial parameters) over varying depths and periods for different rain events.

5.2 METHODS

This section's goal is to examine the quality of roof-harvested rainwater at various depths in a storage tank with a focus on testing the effectiveness of the sedimentation mechanism. This will be achieved by conducting the following sets of experiments:

- Determining the variation of water quality from roof-top harvested rainwater within storage vessels, principally through the mechanism of sedimentation.
- Determining the improvements that can be achieved through the application of household treatment techniques.

In this study, significant microorganisms in rainwater storage tanks are considered to be associated with and housed by particulate matter. Thus, the free phase zone (i.e., the top level) of the rainwater storage tank contains lower levels of bacterial concentration than the sludge zone because of the impact of sedimentation.

5.2.1 Influence of Sedimentation on Microbial and Particle Removal

The set of experiments used in this chapter were conducted at the same location as those in Section 4.1. To collect rainwater from the roof for a variety of rainstorm events, a second gutter and storage container were affixed to the building. The gutter system and storage tanks had a similar layout as in Section 4.1. The storage container of choice had a tight-fitting plastic top and a capacity of 255 L (diameter, D = 600 mm; height, H = 900 mm). In addition to being available and having the capacity to store rainwater for the duration of the measurements, this storage size was chosen because it corresponds to the common size of storage vessels used by the residents in this area (as determined at the questionnaire stage of this study).

The height of the tank is key in this study as it will allow a larger portion of sediment to settle, with sedimentation being an important mechanism in improving water quality. Water from the investigated roof is collected by a gutter, which is 2500 mm long (Figure 5.1). The tank was filled on a day when the rainfall intensity exceeded 10 mm/day, which is commonly encountered in this region. Additional rainfall was redirected using an overflow pipe to the drainage.

Figure 5.1 2500 mm × 200 mm × 100 mm gutter to harvest for sedimentation experiments.

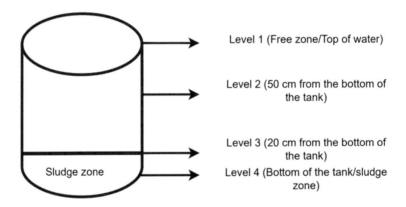

Figure 5.2 Various positions for water removal in the storage tank

Water was collected at multiple depths, i.e., at the top of the tank by gently dipping a pre-sterilised bottle into the storage vessel, and at the bottom of the storage vessel via a siphon pump. Water samples were further collected from intermediate levels by installing small taps at depths shown in Figure 5.2. A small section of rigid pipe 80 mm in length was attached to the inside of each tap to collect water samples from the storage vessel. The storage tank was emptied and washed with sterilised water after each rain event.

The harvested rainwater was completely emptied into another pre-sterilised tank after being collected. This was done to free up the storage tank in case of another rainfall event and to ensure that the particles can be completely mixed before the analysis. This took place immediately after the cessation of the rainfall event. A 5 L pre-sterilised plastic container was used to collect free-fall rainwater samples during the rain event to enable the assessment of the effects of roof contamination during the same rain event (as benchmarked by the ambient atmospheric contamination) (John et al., 2021a).

All the storage vessels, gutters, PVC pipes, and plastic bottles (that were used to collect the roof-harvested rainwater) were pre-washed with sterilised water to prevent external contamination. Furthermore, the sterilised water was tested for bacteria, and the results indicated that there were none present (0 MPN/100 mL of total coliform and *E. coli*). Within 6 hours of collection, the water samples were transported to the laboratory for analysis and kept in an ice-filled refrigerator. After the rainwater was harvested and collected in storage tanks, the lids were thoroughly cleaned. The residual water in the tank was thrown out, and it was cleaned with sterile water following the examination of each rain event. This was done to avoid mixing of old and new rainwater that was obtained externally.

Collections at each level (Figure 5.2) over time were subjected to analyses for suspended particles, microbiological material, and unfractionated as well as fractionated samples. In addition, measurements of turbidity, conductivity, pH, and water temperature were obtained. Tables 5.1 and 5.2 provide a summary of the sampling programme. Four rainfall events (one in the dry season and three in the rainy season) were harvested and analysed for the sedimentation experiment. The measurements were taken from the moment that a rain event occurred till an end until the 20th day (Table 5.1). The fractionated solids and coliform bacteria were analysed at the 2nd, 8th, and 36th hour, whereas the unfractionated solids and coliform bacteria were studied at the 0th, 72nd, 168th, 240th, and 480th hour.

5.2.2 Influence of Household Water Treatment Techniques

As mentioned, rainwater was harvested into a 255 L tank via a 2500 mm gutter, and 5 L of the water sample was collected from the top of the tank (i.e., level 1) after the rainfall event had stopped. This tested level was coincident with the common practice of residents in the study area, who use cups or jugs to collect water from the top of the tank. Five equally sized portions of the 5 L water sample were apportioned for various experiments. The storage tank and water sample containers were cleaned with sterilised water prior to and following each rainfall event's harvest. The microbiological analysis of the sterilised water using the control experiment revealed that neither total coliform nor *E. coli* was present in the sterilised water. This was done to ascertain that the storage tanks were bacteria-free prior to each rainfall harvest. All four rainfall events were also analysed for the

Table 5.1 Programme of samples to be taken at each depth

Time since cessation of rainfall event (Hours)	Turbidity, pH, and conductivity	Suspended solids (fractionated)	Suspended solids (unfractionated)	Coliform bacteria (fractionated)	Coliform bacteria (unfractionated)
0	X		X		X
2	X	X		X	
4	X				
6	X				
8	X	X		X	
10	X				
12	X				
24	X				
36	X	X		X	
48	X				
72	X		X		X
96	X				
120	X				
168	X		X		X
240	X		X		X
360	X				
480	X		X		X
Total	17	3	5	3	5

Table 5.2 Sampling plan for total suspended solids, turbidity, and microbiology

	Solids & turbidity	Colilert test
No. of sample points	6	*6
Repeat experiments	4	4
Duplicate samples	N/A	2
Total No. of samples	24	48

Notes: *Each type of treatment, raw water, and control are taken as a sampling point.

household water treatment technique experiments, i.e., two rainfall events occurred during the rainy season and the other two during the dry season. The household water treatment techniques investigated in this study include:

a) Alum treatment,
b) Boiling treatment,
c) Chlorination treatment, and
d) Combined chlorination and alum treatment.

The first three household water treatment techniques were considered because these are the techniques used by the residents of the study area. The fourth technique was introduced to further explore its effect on the quality of stored rainwater. The definitions of each treatment technique will be fully detailed as follows.

Alum: The particle separation process is improved by the addition of alum to the water. The processes of sedimentation and filtration are used because the treatment is not a stand-alone procedure to clean the water. Alum was bought and applied according to the manufacturer's instructions. This required that 12 g of the powdered alum was weighed, mixed, and dissolved into a litre of raw water sample. The water sample was mixed rapidly (at 110 rpm) and slowly (at 40 rpm) for 3 and 25 minutes, respectively, before being left to settle for 1 hour. The water sample was then examined for TSS, turbidity, total coliform, and *E. coli*.

Boiling: This method involved boiling 1 L of rainwater for 10 minutes at a temperature of 100 °C (Spinks et al., 2006). Measurements of the physical and microbiological parameters were taken before and after boiling to investigate their differences.

Chlorination: Sodium hypochlorite, a common source of chlorine, was utilised in this treatment. In its application, 15 g of powdered material was weighed and dissolved in 1 L of sampled rainwater (as advised by the manufacturer's guidelines) before the physical and microbial parameters were analysed.

Combination of chlorination and alum: A combination of alum and sodium hypochlorite was applied to the water to investigate this combined treatment's impact. 12 g of the powdered alum and 15 g of powdered sodium hypochlorite were weighed and dissolved into 1 L of the raw water sample.

Table 5.3 Rain event characteristics

Rain events	Total storm depth (mm)	Dry antecedent days (days)	Rainfall intensity (mm/h)
1	24.6	13	26.2
2	16.4	3	12.8
3	19.1	8	21.4
4	17.5	5	19.3

The water sample was mixed quickly and slowly before being left for an hour to settle. The physical and microbial parameters were measured before and after the application of the powders.

To investigate the effectiveness of each treatment, the water samples were examined for total coliforms, *E. coli*, suspended particles, and turbidity both in the raw sample and following each treatment. The outcomes of these experiments will aid in supplying the Quantitative Microbial Risk Assessment model with useful information, i.e., on the approximate amount of bacteria ingested by the locals and the outcome of each group (including persons utilising each household water treatment technique). Table 5.2 sets out the expected sampling plan.

5.3 RESULTS AND DISCUSSION

Two of the household water treatment technique experiments were conducted during the dry season, and the other two were conducted during the rainy season; for the sedimentation experiment, one was done towards the late dry season and the remaining three were done in the rainy season. The characteristics of the rain events used for the sedimentation experiments are presented in Table 5.3.

5.3.1 Physical Parameters from Sedimentation Experiments

As was previously mentioned, pH, turbidity, TSS, and water temperature are the key physical factors examined in this study. These parameters were examined since one of the goals of this experimental section was to understand how the sedimentation process interacts or associates with those physical factors. The findings indicated that the water temperature in the storage tank remained almost constant for the different positions in the tank at the same moment.

pH

The level of acidity or alkalinity in water is indicated by its pH level. It can also be referred to as a measurement of water's hydrogen ion concentration. Water is alkaline when it has a pH above 7, and it is acidic when the pH is below 7. Extreme pH values have a significant impact on the health of the water drinker. Low-pH water typically tastes bad and is hence unpleasant to drink. Due to the corrosive effects of low pH levels, drinking water with pH levels below 4 can cause gastrointestinal discomfort and long-term damage to skin and organ linings, while drinking water with a pH level above 11 can irritate the skin, eyes, and mucous membranes (WHO, 2011; Achadu et al., 2013). The World Health Organization recommends a guideline limit for drinking water's pH between 6.5 and 8.5 (WHO, 2011).

The results of the series of tests conducted for the four harvested rainfall events (in Table 5.4) revealed a consistent pattern for the pH at four levels of the storage vessel (Figures 5.3–5.6). The findings indicated that during the first day, the pH value in each of the four levels remained nearly constant. The top levels (1 and 2) experienced an increase in pH values after the first day, while levels 3 and 4 experienced a decrease (see Figure 5.2 for a description of each level). The pH fluctuations in these levels are a result of levels 1 and 2 being more aerated than levels 3 and 4. The top 2 levels in the tank had more exposure to oxygen (from air) than the bottom 2 levels. After the investigation, the pH for all events did not fall within the range of the World Health Organization's guideline limits for drinking water. As seen in Figures 5.3–5.6, the first rain harvest's pH is greater on average than the rest. In the figures, the first harvest occurred during the dry season, whereas the remainder of the harvests took place during the rainy season, and this has contributed to the first harvest's relatively higher pH. Despite the increase in pH values in the top two levels, the pH was not deemed to be acceptable according to the WHO guidelines for drinking water.

The relatively higher pH experienced in the first harvest is attributed to a higher accumulation of particulate matter in the atmosphere during the dry season (John et al., 2021). The quality of the water over time may be affected by the alkalinity and acidity of water. The pH value obtained from these experiments correlates with previous rainwater studies which specify that the pH of most rainwater collected around the world is about 5.6 (Efe, 2010; Olowoyo, 2011; Olaoye and Olaniyan, 2012; Chukwuma et al., 2012; Junaid and Agina, 2014). The reaction between rainwater and the atmospheric acid deposited from several activities, i.e., agricultural operations, industries, dusts from unpaved roads, arid areas, and vehicle exhaust fumes, caused the acidity in the harvested rainwater in the study area. To elevate the pH of the water before drinking, it is advised that lime-softening chemicals (such as lime water and soda ash) to be added. These chemicals would help raise the pH to the required range as stated by the World Health Organization guidelines (De Vera et al., 2015; Kaushal et al., 2018).

Conductivity

The conductivity can illustrate the concentration of dissolved solids which have been ionised in a polarised solution. In other words, the electrical

Table 5.4 Quality of free-fall and roof-harvested rainwater for the four different harvest periods

Parameters	1		2		3		4	
	FFRW	RHRW	FFRW	RHRW	FFRW	RHRW	FFRW	RHRW
TC (MPN/100mL)	27.1	83.1	5.3	19.2	13.7	38.4	11.1	34.4
EC 1 (MPN/100mL)	5.3	22.2	1	6.4	4.2	12.4	3.1	8.7
Turbidity (NTU)	12	78	3.2	19.2	5.8	68.3	5.7	61.9
pH	5.9	6.4	5.3	5.6	5.6	5.7	5.6	5.9
Conductivity (µS/cm)	43	27	26	63	53	164	68	179
Colour (PCU)	10	65	5	15	10	60	5	35
TDS (mg/L)	28	43	18	38	19	61	18	64
TSS (mg/L)	51	182	28	85	49	187	42	196

Note: FFRW and RHRW denote free-fall harvested rainwater and roof-harvested rainwater, respectively, while 1, 2, 3, and 4 represent the four different rain harvests as described.

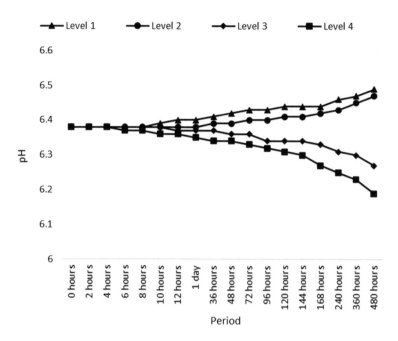

Figure 5.3 pH values at different levels in the storage tank for the 1st harvest

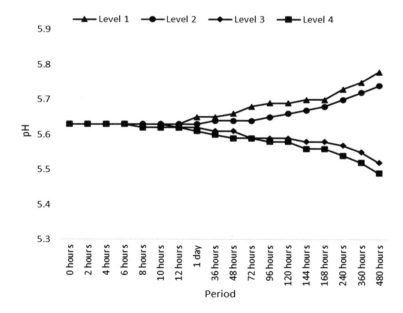

Figure 5.4 pH values at different levels in the storage tank for the 2nd harvest

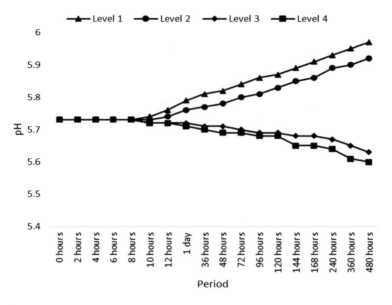

Figure 5.5 pH values at different levels in the storage tank for the 3rd harvest

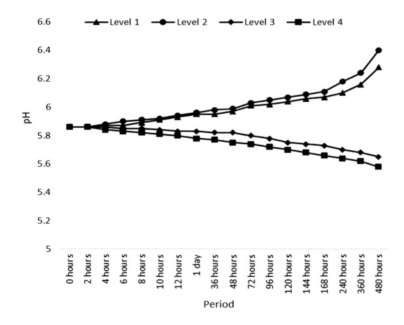

Figure 5.6 pH values at different levels in the storage tank for the 4th harvest

conductivity of water approximates the total quantity of solids dissolved in water (WHO, 2003; Adeniyi and Olabanji, 2005; WHO, 2011). The conductivity values of the harvested stored rainwater at different levels in the storage tank from the cessation of the rain till the 20th day for the four different harvests are shown in Figures 5.7–5.10. The figures' findings demonstrate that the conductivity values of the harvested rainwater did not exceed the World

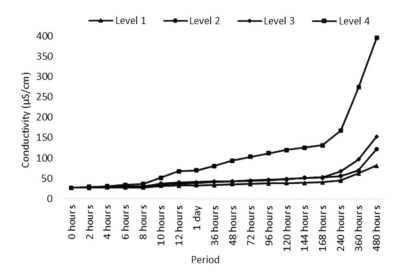

Figure 5.7 Conductivity values at different levels in the storage tank for the 1st harvest

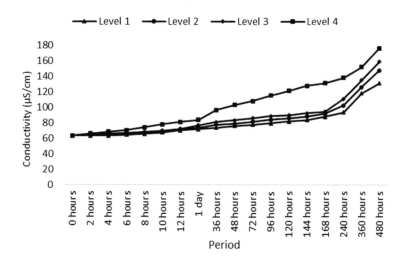

Figure 5.8 Conductivity values at different levels in the storage tank for the 2nd harvest

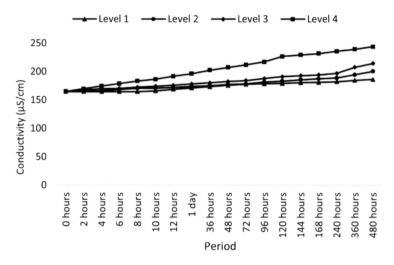

Figure 5.9 Conductivity values at different levels in the storage tank for the 3rd harvest

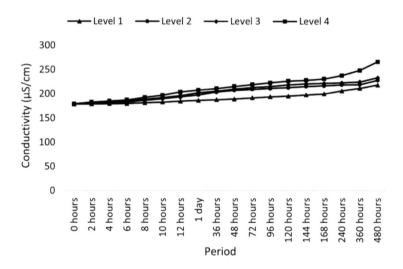

Figure 5.10 Conductivity values at different levels in the storage tank for the 4th harvest

Health Organisation's recommended limits of 1,400 µS/cm (WHO, 2011). From the time of rain stopped until the 20th stored day for all four different harvests, there was a steady rise in the conductivity values of the rainwater that had been collected at the four different described levels in the storage tank. The results, however, showed that level 4 had the highest rate of conductivity increment and level 1 had the lowest rate. The highest conductivity values at level 4 were caused by an increase in total dissolved solids (TDS) at that level. The plots for the four rainfall events showed agreeing increment patterns at different tank levels (Figures 5.7–5.10). The sedimentation process increased the solid content at the bottom of the tank with a resultant increase in conductivity.

Turbidity

Water turbidity is typically caused by colloidal matter, or particles floating in the water that prevent light from passing through. Water becomes cloudy due to suspended particles, which might be organic or inorganic. The organic particles may also comprise various microbes, fats, grease, or animal matter, whereas the inorganic particles may include iron, clay, industrial wastes, manganese, calcium carbonate, sulphur, rock flour, silt, or silica (WHO, 2003; WHO, 2011; Yang et al., 2015). The World Health Organization's recommended limit for turbidity in drinking water is 5 NTU, and the results from the investigated case study revealed turbidity levels above this limit (Table 5.4). Turbidity does have long-term effects on health, even though they are not immediately apparent.

Results from Chapter 4 and research by Characklis et al. (2005) and Krometis et al. (2013) revealed that microbes (such as viruses, bacteria, and protozoa) are typically attached to suspended solids in water. As a result, the presence of suspended solids could enhance microbial contamination in the water. Furthermore, increased turbidity can protect microorganisms from influence of chlorination and stimulate the growth of bacteria, thereby leading to a significant increase in chlorine use in treatment (WHO, 2011). Figures 5.11–5.14 show the turbidity plot for the four harvested rainfalls at different levels in the storage tank. Over 50% of the reduction in turbidity

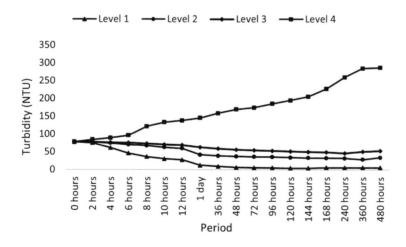

Figure 5.11 Turbidity values at different levels in the storage tanks for the 1st harvest

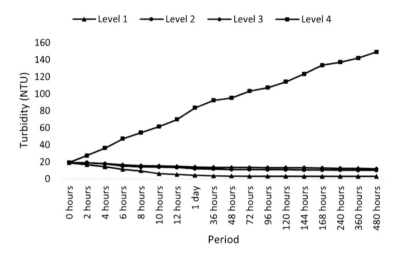

Figure 5.12 Turbidity values at different levels in the storage tanks for the 2nd harvest

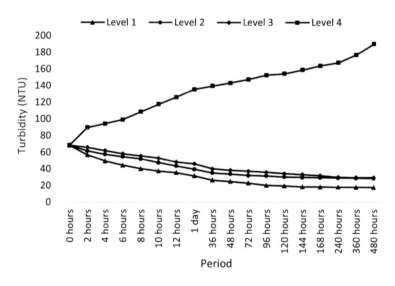

Figure 5.13 Turbidity values at different levels in the storage tanks for the 3rd harvest

was seen within the first 24 hours of harvest for levels 1, 2, and 3 of the storage tank, with a continual reduction in the turbidity values due to particles settling. The rate of settling was highest in level 1 and lowest in level 3, as sedimentation of particles depended on the depth of the storage tank. The turbidity value for level 4 increased steadily throughout the course of the experiment due to settle-able particles from the top to the bottom of the tank.

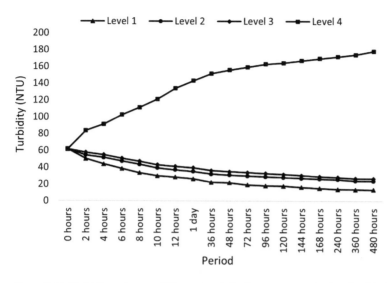

Figure 5.14 Turbidity values at different levels in the storage tanks for the 4th harvest

Unfractionated Total Suspended Solids for Different Periods

The values of TSS for the stored rainwater for 20 days are presented and shown in this section. TSS values decreased for the storage tank's top three levels, but a steady increase in TSS values was seen at the tank's bottom level (Figure 5.15 to 5.18) due to the settling of suspended particles. Level 3 and

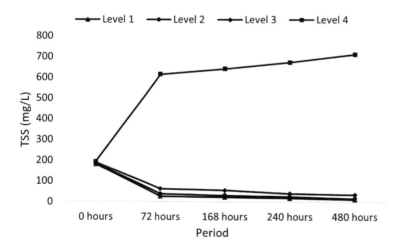

Figure 5.15 Unfractionated total suspended solids (mg/L) for different periods for the 1st harvest

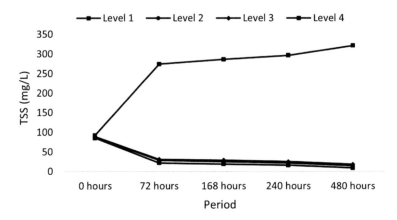

Figure 5.16 Unfractionated total suspended solids (mg/L) for different periods for the 2nd harvest

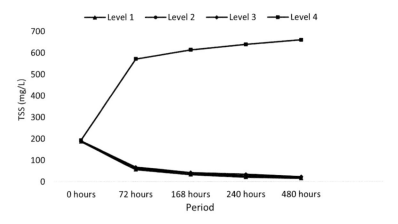

Figure 5.17 Unfractionated total suspended solids (mg/L) for different periods for the 3rd harvest

level 1 had the lowest and highest rates of settling, respectively, due to their depths in the storage tank. In addition, it was observed that more than 80% of sedimentation took place within the first three days following the end of the rain, whilst the last two days had the least amount of particle settling. Experimental results from the four rainfall events produced similar patterns (Figures 5.15–5.18), which indicates the consistency of this study's measurements. These findings also illustrate that the sedimentation process is crucial for the enhancement of the water quality.

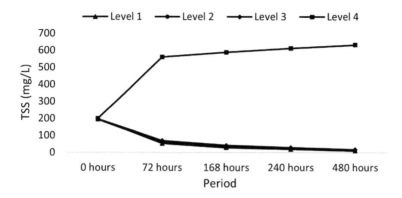

Figure 5.18 Unfractionated total suspended solids (mg/L) for different periods for the 4th harvest

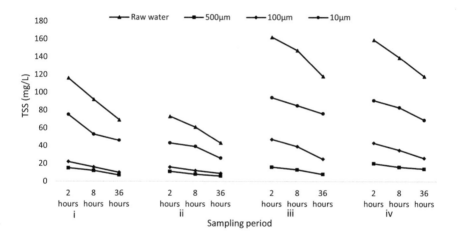

Figure 5.19 Fractionation of TSS (mg/L) at level 1 for the four rain harvests

Fractionation of Total Suspended Solids after 2 hours, 8 hours, and 36 hours

The size range of suspended solids from four varying depths in the storage tank has been investigated and presented in Figures 5.19–5.22, while Table 5.5 presents the percentage of particles retained in the four levels of the storage tank for the four rainfall events over the first 36 hours of harvest. In the top three levels of the storage vessel, TSS consistently decreased following each rainfall event, but it increased at the bottom due to TSS settling. Table 5.5 was obtained by dividing the mass of particles retained on each filter by the total mass collected using the 1.5 μm filter paper. The results from

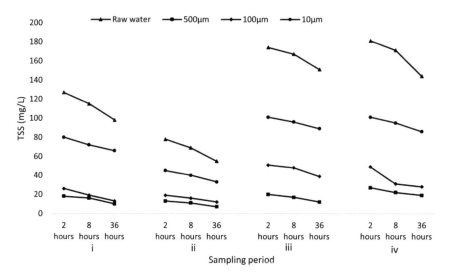

Figure 5.20 Fractionation of TSS (mg/L) at level 2 for the four rain harvests

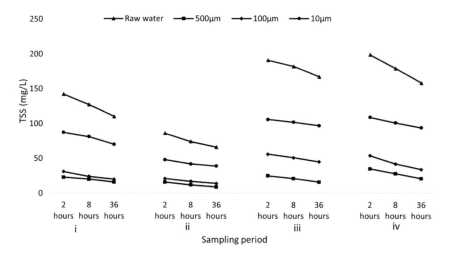

Figure 5.21 Fractionation of TSS (mg/L) at level 3 for the four rain harvests

fractionation also showed that majority of the particle sizes ranged between 100 μm and 10 μm, whereas the least common particle size was about 500 μm. The percentage of solids (from the first rain event) retained in the 500 μm mesh filter during three different periods (i.e., 2 hours, 8 hours, and 36 hours) in the four levels ranged from 10.1% to 16.2%, whilst the percentage retained by the 10 μm filter ranged from 57.6% to 72.2%. Comparatively,

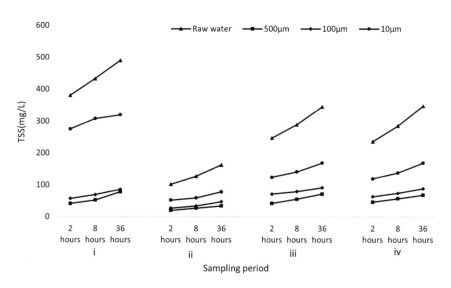

Figure 5.22 Fractionation of TSS (mg/L) at level 4 for the four rain harvests

during the second rain event, the percentage of TSS retained by the 10 μm filter ranged from 47.8% to 63.9%, whereas the proportion deposited in the 500 μm mesh filter in the four levels ranged from 12.7% to 21.3%. The third and fourth rain events both showed similar trends. The fractionation outcomes, however, indicated that not all the solids in the initial samples had been removed, and this is due to some of the solids being smaller than a 10 μm filter. For the four rainfall events, the percentage range of these unaccounted solids was 1–12, 1.2–6, 0.5–13.4, and 1.2–7.6 (Table 5.5). The outcomes of these experiments further evidence that the sedimentation process significantly improved the rainwater's quality in terms of suspended solids. This is because over 50% of the TSS values were reduced over a three-day period.

5.3.2 Microbial Parameters from the Sedimentation Experiments

This section presents the enumerated bacteria that were discovered through the study of roof-harvested rainwater as well as the outcomes of the serial filtration of the enumerated microorganisms. *E. coli* and total coliform were among the microorganisms analysed in this study. Every microbiological test was repeated twice. Some researchers, such as Mahmood (1987) and Gray (2010) examined the impact of sedimentation in water treatment facilities; however, there is limited study on the impact of sedimentation in rainwater storage tanks, especially in Nigeria (John et al., 2021b, 2021c). The World Health Organization's recommended limit for microbes in drinking water is

Table 5.5 Percentage of particles retained in the four levels at different periods for four rain events

		2 hours				8 hours				36 hours			
		level 1	level 2	level 3	level 4	level 1	level 2	level 3	level 4	level 1	level 2	level 3	level 4
1st rain event	500μm	12.9	14.2	16.2	10.8	13	13.9	15.7	12	10.1	10.2	14.5	15.9
	100μm	19	20.5	14.8	15	17.4	16.5	18.9	15.9	14.5	13.3	18.2	17.6
	10μm	64.7	63	61.3	72.2	57.6	62.6	63.8	71.1	66.7	67.3	63.6	65.3
	<10μm	3.4	2.3	7.7	2	12	7	1.6	1	8.7	9.2	3.7	1.2
2nd rain event	500μm	15.1	16.7	18.6	19.6	13.1	15.9	16.2	21.3	14	12.7	13.6	20.9
	100μm	21.9	24.3	24.4	26.5	19.7	23.1	23	26.8	20.9	21.8	21.2	28.8
	10μm	58.9	57.7	55.8	51	63.9	58	56.8	47.9	60.4	60	59.2	47.8
	<10μm	4.1	1.3	1.2	2.9	3.3	3	4	4	4.7	5.5	6	2.5
3rd rain event	500μm	12.6	14.9	17.6	19.5	11.5	12.9	15.6	20	11.9	13.2	13.3	19.6
	100μm	27	27.1	27.1	26.7	25.2	18.1	23.5	26	21.2	19.4	21.5	25.4
	10μm	57.2	55.8	54.8	50.4	59.7	55.6	56.4	48.4	58.5	59.7	59.5	48.7
	<10μm	3.2	2.2	0.5	3.4	3.6	13.4	4.5	5.6	8.4	7.7	5.7	6.3
4th rain event	500μm	9.9	11.5	13.1	17	8.8	10.2	11.5	19	6.8	7.9	9.6	20.6
	100μm	29	29.3	29.3	28.7	26.5	28.7	28	27.3	21.2	25.8	26.4	26.4
	10μm	58	58	55.5	50.2	57.8	57.5	56	48.9	64.4	58.9	58.1	49
	<10μm	3.1	1.2	2.1	4.1	6.9	3.6	4.5	4.8	7.6	7.4	5.9	4

0 MPN/100 mL, and the analysis of unfractionated bacteria (i.e., total coliform and *E. coli*) in Table 5.4 revealed that the quality of the stored rainwater is generally unsatisfactory for consumption.

Unfractionated Total Coliform Bacteria and E. coli for Different Periods

The plots of the enumerated total coliform and *E. coli* bacteria for the four harvests of rainwater that was stored until the 20th day are shown in Figures 5.23–5.26. The duplicate and diluted test results displayed patterns that matched the primary result outcomes. The results showed that over 50% of the total coliform and *E. coli* bacteria settled from the top level of the storage tank over the first three days for the four different rainfall harvests, resulting in a steady increase in the total coliform near the bottom of the storage tank (Figures 5.23–5.26). The settling of sediment resulted in the removal of its attached bacteria to the bottom of the tank, causing this outcome. This further highlights the importance of the sedimentation process (i.e., the settling of particles) in the storage tank.

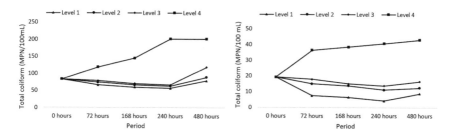

Figure 5.23 Undiluted total coliform bacteria (MPN/100mL) for the 1st and 2nd harvest, respectively

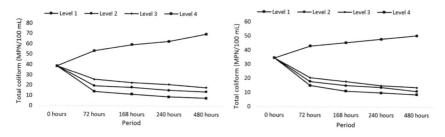

Figure 5.24 Undiluted total coliform bacteria (MPN/100mL) for the 3rd and 4th harvest, respectively

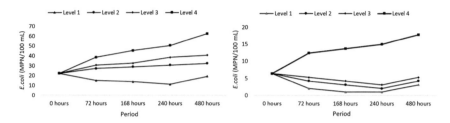

Figure 5.25 Undiluted *E. coli* (MPN/100mL) for the 1st and 2nd harvest, respectively

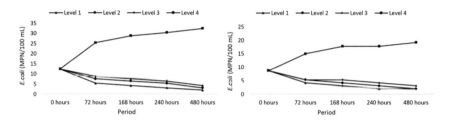

Figure 5.26 Undiluted E. coli (MPN/100mL) for the 3rd and 4th harvest, respectively

Additionally, it was noted that total coliform and *E. coli* values rose in level 1 from the 240th hour to the 480th hour, especially in the first and second rainfall events (Figures 5.23 and 5.25). This is due to bacteria re-growth which occurred because of a viable environment. According to Hill's (2006) investigation into the bacterial activity in the harvested and stored rainwater, its re-growth in the rainwater storage tank was caused by the presence of sufficient nutrient substrate for the bacteria to feed on. The presence of organic materials, enough time for the bacteria to multiply, a warm environment, and the chemical makeup of the water are key elements that promote the growth of bacteria (Hill, 2006).

Fractionation of Total Coliform and E. coli after 2 hours, 8 hours, and 36 hours

This section presents the total coliform and *E. coli* counts that were determined after rainwater was stored for certain periods (i.e., after 2 hours, 8 hours, and 36 hours). Additionally, the findings of the serial filtration of the counted total coliform and *E. coli* bacteria are reported in this section. All microbiological tests were done in duplicate. The findings from the series of experiments revealed that a continuous increase in the bacterial counts (for both total coliform and *E. coli*) was observed in the bottom three levels (i.e., levels 2, 3, and 4) of the tank, while level 1 experienced a reduction. The settling of particles is the reason for the change in the number of

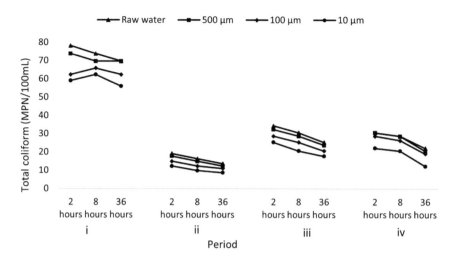

Figure 5.27 Fractionation of Total coliform (mg/L) at level 1 for the four rain harvests

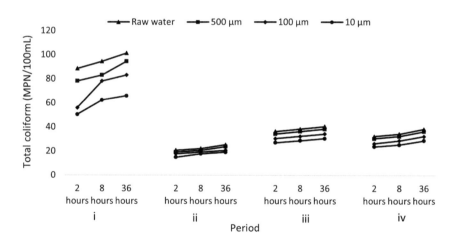

Figure 5.28 Fractionation of total coliform (mg/L) at level 2 for the four rain harvests

microorganisms at different levels of the storage tank. Levels 2 and 4 had the least and most enumerated bacteria, respectively, among the three levels that saw an increase in bacteria (Figures 5.27–5.33). This demonstrates that the quality of rainwater at top level was improved by the sedimentation.

Figures 5.27–5.30 and Figures 5.31–5.34, respectively, display the fractionated total coliform and *E. coli* plots for the roof-harvested and stored rainwater tanks for different events. Tables 5.6 and 5.7, respectively, show the percentage of bacteria removed by various filter sizes in the four different

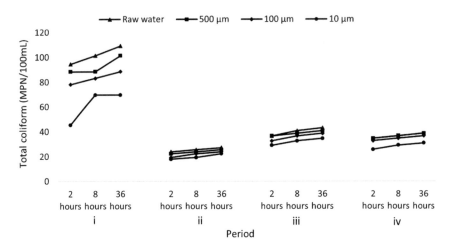

Figure 5.29 Fractionation of total coliform (mg/L) at level 3 for the four rain harvests

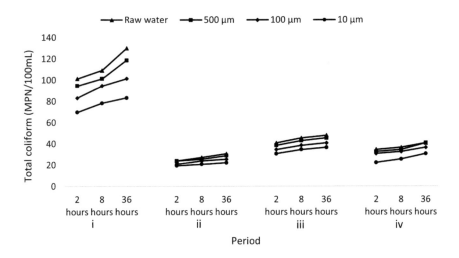

Figure 5.30 Fractionation of Total coliform (mg/L) at level 4 for the four rain harvests

levels of the storage tank over the first 36 hours post-harvest. The result of serial filtration revealed that the majority of bacteria was removed by the 10 μm filter size, whereas the least was removed by the 500 μm mesh filter size. The percentage of total coliform retained by the 10 μm filter in the four rainfall events ranged from 4.2 to 32.9, 5.9 to 17.5, 8.8 to 15.4, and 8 to 30.6 respectively, while the percentage removed by the 100 μm filter ranged from 5.1 to 25.1, 5.9 to 15.9, 4.9 to 12.2, and 5.2 to 13 respectively (Table 5.6). The percentage of *E. coli* counts in the 10 μm filter for the four rain

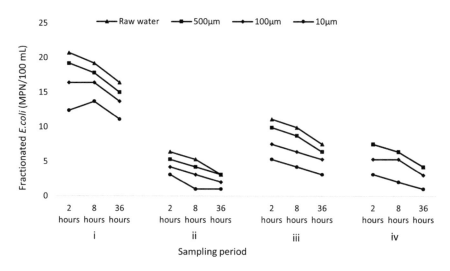

Figure 5.31 Fractionation of E. coli (MPN/100 mL) at level 1 for the four rain harvests

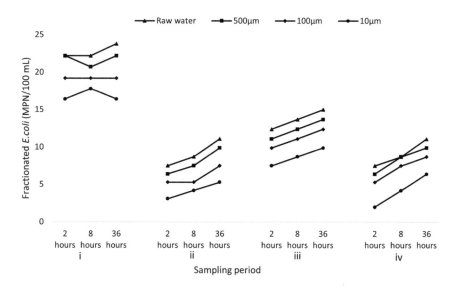

Figure 5.32 Fractionation of E. coli (MPN/100 mL) at level 2 for the four rain harvests

events ranged from 4.9 to 19.3, 12.6 to 39.6, 8 to 29.3, and 20.7 to 51.6 respectively, while the retainment by the 100 μm filter ranged from 5.6 to 13.5, 10.8 to 35.5, 8.7 to 23.2, and 10.5 to 29.3 respectively (Table 5.7). For the 500 μm mesh filter test, the percentage of total coliform removed by the four rain events ranged from 0 to 12.6, 0 to 9.5, 0 to 6.3, and 0 to 6.8,

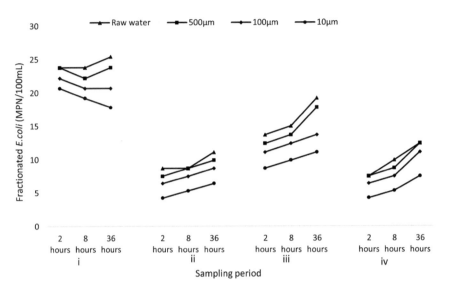

Figure 5.33 Fractionation of *E. coli* (MPN/100 mL) at level 3 for the four rain harvests

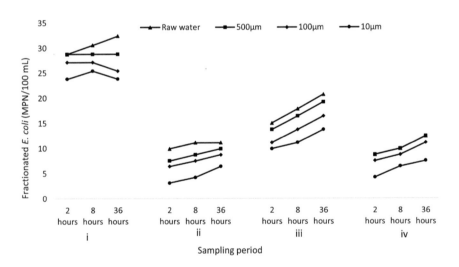

Figure 5.34 Fractionation of *E. coli* (MPN/100 mL) at level 4 for the four rain harvests

respectively, while the retained *E. coli* ranged from 0 to 11.1, 0 to 24.2, 7.2 to 14.7, and 0 to 14.7, respectively.

The analysis of the results above showed that little or no bacteria (i.e., total coliform and *E. coli*) was removed by the 500 μm mesh filter, while most bacteria was retained by the 10 μm filter. The removal of bacteria by the filter paper can be attributed to some bacteria being attached to the suspended solids, which is supported by findings from Krometis et al. (2013) and John et al. (2021b), who had previously reported that settle-able bacteria are attached to suspended solids. Therefore, it can be concluded that the collection of solids by the filters during the serial filtration led to the reduction in the enumerated bacteria. The analysis of the results further proved that not all the bacteria in the raw samples were accounted after the serial filtration due to some free phase bacteria that was not attached to any solids. Besides, it could be attributed to some solids being smaller than the 10 μm filter. In the four investigated rainfall events, the percentage of the unaccounted total coliform ranged from 47.9 to 84.6, 60.4 to 81.9, 67.6 to 80.2, and 55.8 to 79.7, respectively (Table 5.6), while the percentage of the unaccounted *E. coli* ranged from 60 to 87, 18.8 to 60.9, 41.3 to 69.3, and 23.8 to 64.7, respectively (Table 5.7).

Finally, the results from this chapter show that the sedimentation process significantly contributes to the improvement of rainwater quality with regards to settle-able bacteria and the physical parameters. Results showed about 30% reduction in the enumerated bacteria in the level 1 of the storage tank, while over 75% reduction was observed in turbidity and TSS in that top level. On analysis of the fractionation results, single-celled bacteria made up 70% of the microbial load, and these bacteria can easily be killed using heat treatment or disinfection, which leads to the recommendation that harvested rainwater should be treated to improve its physical and microbial parameters.

Among the physical parameters, the pH and turbidity values of the roof-harvested rainwater exceeded the World Health Organization's recommended limit for drinking water, whereas for microbiological parameters, the total coliform and *E. coli* results were over the WHO drinking water guideline limit. As a result, the harvested rainwater is not safe for drinking except with the application of some form of disinfectant. The contamination is due to pollution from dust/particulate matter and animal faeces deposition on the roof. These pollutions are typically influential on rainwater quality during extended dry antecedent periods, as compared to the rainy season, where short dry antecedent days before rainfall events occur.

The sedimentation stage in the water treatment process and in a rainwater storage tank is a low-cost and simple method which can reduce turbidity and lay the foundation for subsequent treatment (Gregory and Edzwald, 2010). For example, it was used to lower the concentration of suspended particles before the application of coagulant in order to lessen the amount of coagulating chemicals required. Sedimentation is one of numerous techniques that can be used before filtering treatment; other choices also include dissolved air flotation. These procedures for separating particles from liquids are frequently referred to as clarifying processes (Gregory and Edzwald, 2010). The sedimentation process in this study helped to reduce over 70% of the TSS within the first 36 hours, while over 50% of *E. coli* settled from the top over the first three days. Water turbidity can be reduced naturally by sedimentation. However, this method is not always successful in decreasing microbial contamination, and point-of-use treatment is required instead.

Table 5.6 Percentage of total coliform removed by different filter sizes for the four rain events

		2 hours				8 hours				36 hours			
		level 1	level 2	level 3	level 4	level 1	level 2	level 3	level 4	level 1	level 2	level 3	level 4
1st rain event	500 μm	5.6	11.7	6.3	6.7	5.6	12.1	12.6	7.2	0	6.7	7.1	8.8
	100 μm	14.6	25.1	10.9	11.3	5.1	5.2	5.4	6.2	10.5	11.3	11.8	13.2
	10 μm	4.2	6.3	32.9	13.2	4.7	16.7	13.2	14.9	9.2	17	17.2	14
	% removed	24.4	43.1	52.1	31.2	15.4	34	31.2	28.3	19.7	35	36.1	36
2nd rain event	500 μm	7.3	7.2	6.7	0	8.5	6.8	6.3	6.3	9.5	6.3	6.3	5.9
	100 μm	14.6	6.7	12.6	13	15.9	6.8	6.3	5.9	9.5	12.2	5.9	11.1
	10 μm	13.5	13.5	5.9	6.3	15.2	6.3	11.8	11.4	17.5	5.9	5.9	10.5
	% removed	35.4	27.4	25.2	19.3	39.6	19.9	24.4	23.6	36.5	24.4	18.1	27.5
3rd rain event	500 μm	5.8	5.5	0	5.4	5.9	5.2	5.4	5.3	6.3	5.4	5.4	5.2
	100 μm	10.5	10.4	11	9.9	11.1	10.4	4.9	9.9	12.2	9.9	5.1	9.8
	10 μm	9.9	9.6	9.9	9.3	15.4	9.4	9.9	8.8	11.4	9.4	9.3	8.8
	% removed	26.2	25.5	20.9	24.6	32.4	25	20.2	24	29.9	24.7	19.8	23.8
4th rain event	500 μm	0	5.6	5.8	5.8	0	5.8	0	5.5	6.8	5.2	0	0
	100 μm	5.9	13	0	5.2	8.3	10.5	5.5	5.5	6.8	10.4	5.2	10.3
	10 μm	21.6	8	20.3	24.4	19.8	9.9	15.4	19.2	30.6	9.4	15.1	14.3
	% removed	27.5	26.6	26.1	35.4	28.1	26.2	20.9	30.2	44.2	25	20.3	24.6

Table 5.7 Percentage of E. coli removed by different filter sizes for the four rain events

		2 hours				8 hours				36 hours			
		level 1	level 2	level 3	level 4	level 1	level 2	level 3	level 4	level 1	level 2	level 3	level 4
1st rain event	500 μm	7.2	0	0	0	7.3	6.8	6.7	5.9	8.5	6.7	6.3	11.1
	100 μm	13.5	13.5	6.7	5.9	7.3	6.8	6.3	5.6	7.9	12.6	12.2	10.5
	10 μm	19.3	6.3	6.3	11.5	14.1	6.3	6.3	5.6	15.9	11.8	11.4	4.9
	% removed	40	19.8	13	17.4	28.7	19.9	19.3	17.1	32.3	31.1	29.9	26.5
2nd rain event	500 μm	17.2	14.7	13.8	24.2	20.8	13.8	0	21.6	0	10.8	10.8	10.8
	100 μm	17.2	14.7	12.6	11.1	20.8	25.3	13.8	10.8	35.5	21.6	10.8	10.8
	10 μm	17.2	29.3	25.3	33.3	39.6	12.6	25.3	29.7	32.3	19.8	20.7	20.7
	% removed	51.6	58.7	51.7	68.6	81.2	51.7	39.1	62.1	67.8	52.2	42.3	42.3
3rd rain event	500 μm	10.8	10.5	9.5	8.7	12.1	9.5	8.7	7.9	14.7	8.7	7.9	7.2
	100 μm	21.6	9.7	9.5	17.3	23.2	9.5	8.7	15.2	14.7	8.7	15.2	13.5
	10 μm	19.8	19.4	17.5	8	22.2	17.5	16.7	14.6	29.3	16.7	14.6	13
	% removed	52.2	39.6	36.5	34	57.5	36.5	34.1	37.7	58.7	34.1	37.7	33.7
4th rain event	500 μm	0	14.7	0	0	0	0	12.1	0	0	10.8	0	0
	100 μm	29.3	14.7	14.7	13.8	17.2	13.8	12.1	12.1	28.6	10.8	10.5	10.5
	10 μm	29.3	44	29.3	37.9	51.6	37.9	22.2	23.2	47.6	20.7	29	29
	% removed	58.6	73.4	44	51.7	68.8	51.7	46.4	35.3	76.2	42.3	39.5	39.5

Despite the positive impacts of the sedimentation process, the rainwater quality remains unsafe for drinking as the measured microbiological parameters exceeded the World Health Organization's guideline limit for drinking water, as explored in previous paragraphs. The quality of roof-harvested rainwater can be improved through the practice of first flush, which is a process whereby the initial surface runoff of the rainfall is allowed to flow without harvesting. This practice usually helps to wash the roof catchment, pipe system, and gutter, especially after a long period of dry antecedent days. The next chapter will investigate the impact of the first flush practice on improving rainwater quality. Household water treatment techniques are also used to clean rainwater before drinking. These methods include solar disinfection (SODIS), disinfection with chemicals (i.e. chlorination), slow sand filtration, and boiling. Though the impacts of SODIS and slow sand filtration were not measured in this study, they can be briefly explained. The SODIS technique is a low-cost water treatment technology for tropical areas with high sunlight intensity. The technique can effectively cleanse small amounts of water with low turbidity (<30 NTU), such as <10 litres, and it is ideal for areas of high sunlight exposure. It is highly effective against bacteria, viruses, and protozoa, and it is easy to handle with little change to the taste of the water. However, it has a waiting period of 6–12 hours, and it does not remove suspended particles or dissolved compounds (Clasen, 2006). The slow sand filtration technique works through a combination of biological and physical processes. It can efficiently eliminate turbidity (cloudiness) and bacteria in a single treatment step. A slow sand filter is made up of vertical layers of components. The filter is made up of a tank, a bed of fine sand, a layer of gravel to support the sand, a network of underdrains to catch the filtered water, and a flow regulator to regulate the filtration rate (CDC, 2019). The merits and demerits of chlorination and boiling have been discussed in the above section.

5.3.3 Statistical Analysis of the Measured Parameters

Using the unfractionated sample data of turbidity, TSS, and *E. coli* over 0, 72, 168, 240, and 480 hours, the Pearson's correlation[2] was further performed. The correlation matrix between turbidity and *E. coli* between the four storage levels in all four rain occurrences over the five specified durations is presented in Table 5.8. The findings showed a generally high correlation for all rain occurrences across all levels, with levels 3 and 4 showing the lowest (0.68) and highest (0.996) correlation, respectively. In line with this study, numerous additional investigations have also found a strong association between turbidity and the number of bacteria present (Irvine et al., 2002; Huey & Meyer 2010; Lawrence, 2012; John et al., 2021b).

In addition, Table 5.9 shows a correlation matrix between turbidity and solids (TSS) for all levels in all rainfall events over the same period (i.e.,

Table 5.8 Pearson's correlation between *E. coli* and turbidity for four rain events in four storage tank levels

	Level 1		Level 2		Level 3		Level 4	
	E. coli	*Turbidity*	*E. coli*	*Turbidity*	*E. coli*	*Turbidity*	*E. coli*	*Turbidity*
1st event *E. coli*	1.00		1.00		1.00		1.00	
Turbidity	0.75	1.00	0.88	1.00	0.83	1.00	0.98	1.00
2nd event *E. coli*	1.00		1.00		1.00		1.00	
Turbidity	0.92	1.00	0.84	1.00	0.68	1.00	0.97	1.00
3rd event *E. coli*	1.00		1.00		1.00		1.00	
Turbidity	0.97	1.00	0.91	1.00	0.91	1.00	1.00	1.00
4th event *E. coli*	1.00		1.00		1.00		1.00	
Turbidity	0.97	1.00	0.93	1.00	0.95	1.00	0.97	1.00

Table 5.9 Pearson's correlation between turbidity and solids for four rain events in four storage tank levels

		Level 1		Level 2		Level 3		Level 4	
		Turbidity	Solids	Turbidity	Solids	Turbidity	Solids	Turbidity	Solids
1st event	Turbidity	1		1		1		1	
	Solids	0.997	1	0.9936	1	0.9678	1	0.9341	1
2nd event	Turbidity	1		1		1		1	
	Solids	0.988	1	0.9939	1	0.9996	1	0.9836	1
3rd event	Turbidity	1		1		1		1	
	Solids	0.991	1	0.9886	1	0.9975	1	0.9842	1
4th event	Turbidity	1		1		1		1	
	Solids	0.995	1	0.9947	1	0.9967	1	0.9995	1

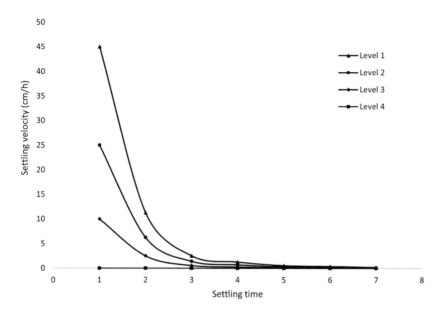

Figure 5.35 Sedimentation velocity of the particles at different settling time periods (settling time 1–7 denotes 2, 8, 36, 72, 168, 240, and 480 hours, respectively)

over five investigated times). The findings have consistently proven high correlation (minimum of 0.9) within all four levels, with similar conclusions also proposed by previous studies (Draper & Smith, 1981; Ferguson, 1986; Davies- Colley & Smith, 2001; Embry, 2001). The majority of the correlation values were over 90%, and these high correlations are the key evidence that significant proportion of *E. coli* fraction is both settle-able and attached to the solids during sedimentation.

Figure 5.35 describes the calculated settling velocities at the four levels in the storage tank during different periods of 2, 8, 36, 72, 168, 240, and 480 hours. The minimum and peak particle settling velocities were observed at level 4 and 1, respectively. The settling rate of particles in a storage tank is dependent on the depth of the tank and the size of the particles present. Conclusively from Figure 5.35, over 70% settlement of TSS was observed within a period of 36 hours when unfractionated water samples were analysed.

5.3.4 Effect of Different Household Water Treatment Techniques (HHTTs)

The HHTTs employed in this investigation are the use of boiling, alum, chlorine, and alum and chlorine combined, as previously mentioned. Table 5.10

Table 5.10 Effect of the different HHTTs on the harvested rainwater for the four rain events

		TC (MPN/100 mL)	EC (MPN/100 mL)	Turbidity (NTU)	TSS (mg/L)
1st rain event	Raw water	69.7	11.1	56.39	358
	After alum	53.1	5.3	4.9	19.5
	After chlorine	0	0	42.7	312
	After boiling	0	0	37.4	274
	After alum + chlorine	0	0	2.89	11.8
2nd rain event	Raw water	22.2	0	17.4	149
	After alum	19.2	0	4.9	10.36
	After chlorine	0	0	15.27	125
	After boiling	0	0	14.3	143
	After alum + chlorine	0	0	1.85	9.41
3rd rain event	Raw water	19.2	6.4	19.86	155
	After alum	19.2	11.1	2.53	12.37
	After chlorine	0	0	12.86	119
	After boiling	0	0	11.3	148
	After alum + chlorine	0	0	1.73	11
4th rain event	Raw water	38.4	12.4	68.27	187
	After alum	34.4	9.9	6.3	24
	After chlorine	0	0	61.83	163
	After boiling	0	0	64.52	152
	After alum + chlorine	0	0	3.17	18

demonstrates the impact of different HHTTs used by households that collect and consume rainwater, where the water samples from level 1 were subjected to the HHTTs. Both total coliform and *E. coli* were absent, according to the analysis of the sterilised water used as the control experiment. This was done to ensure that the storage tanks were bacteria free prior to each rainfall harvest. For the HHTT experiment, four rainfall events were analysed, with the first two taking place during the dry season and the final two during the rainy season (Table 5.10).

Effect of HHTTs on Bacteria

Boiling, chlorine, and the combination of alum and chlorine were shown to be the most effective techniques to eliminate total coliform and *E. coli*, while alum alone produced a very limited effect on bacteria decrease (Table 5.10). Alum is a coagulant, not a disinfectant, and thus it has a minimal impact on the bacteria count. Both boiling and chlorination are primarily disinfectants in the study area and have their merits and limitations.

Boiling is primarily used in rural areas of less economically developed countries. It is frequently used in urgent situations, when other or more refined techniques of disinfection are not easily accessible. Before chlorination became widely popular, boiling was the primary approach (Clasen et al., 2008; Psutka et al., 2011; UNEP, 2012). When electricity or fossil fuels are not available or there is a requirement to treat large amounts of water quickly, boiling presents a problem. Furthermore, the study conducted by Spinks et al. (2006) recommends that (1) the temperature range between 55°C and 65°C is decisive for effective elimination of pathogenic bacteria and (2) drinking hot water systems should operate at a minimum temperature of 60 °C. The results from this study support these recommendations. However, the overall cost of boiling large volumes of water before drinking in rural areas might not be a cost-effective option due to various reasons, e.g., the availability and cost of electricity.

Compared to other HHTTs, chlorination is less expensive and considerably simpler to carry out (Clasen et al., 2008; Psutka et al., 2011; UNEP, 2012). When used properly, chlorine is an efficient disinfectant, but its usage in a solution or gaseous state poses safety concerns, and thus all users should be made aware of these risks. Additionally, since chlorine is so reactive, it can interact with a variety of other elements and compounds found in water, including metals, ammonia, and hydrogen sulphides. The pH and temperature of the water might lead to a decrease in the chlorine concentration, leaving residual levels of chlorine for disinfection. Contrarily, excessive chlorination may result in the development of halogenated hydrocarbons (such as trichloromethanes and tetracloromethane), which have been linked to cancer. For this reason, it is crucial that the correct amount of chlorine is added to the appropriate volume of water (Clasen et al., 2008; Psutka et al., 2011; UNEP, 2012).

Effect of HHTTs on Sedimentation

The results from applying the various HHTTs demonstrate that boiling and chlorination have the least impact on turbidity and TSS, whereas alum and the combination of alum and chlorine had the greatest impact (Table 5.10). Alum is a typical coagulant which will combine smaller particles (by neutralising the electrical charges of water) and causes the particles to lump together, thus ensuring easy settling and filtration. The reduction in the turbidity and TSS values can be attributed to the coagulant nature of alum. The use of alum as the HHTT would only help improve the clarity of the water and reduce the settle-able bacteria that are attached to the particles.

Since viruses, bacteria, and protozoa (microbes) are typically connected to solids (as was established in Chapter 4 and Section 5.3.2, the removal or reduction of turbidity by filtration will significantly reduce the microbiological pollution of water. Furthermore, the removal of the solids by coagulation (i.e., the use of alum) and sedimentation (in the storage tanks) before filtration can path a healthy pre-condition for chlorination. It is therefore imperative to achieve lower turbidity by filtration prior to disinfection of the water to increase the effectiveness of chlorination and limit its excessive use. The combination of chlorine and alum is the most effective of the investigated HHTTs because of the synergy effect of alum acting as a coagulant and chlorine as a disinfectant. Alum helps to settle suspended solids and the attached bacteria in water, while the addition of chlorine to water destroys disease-causing pathogens (such as protozoa and bacteria) because it is an effective disinfectant.

5.4 CONCLUSION

This chapter examined the impact of sedimentation and the different household treatment techniques on the quality of roof-harvested and stored rainwater. The conclusions based on the results of the measurements are as follows:

- Rainwater collected from free fall is of higher quality than rainwater harvested from roofs. However, both types of harvested rainwater were poor in quality and did not comply with the World Health Organization's drinking water quality guidelines. Thus, it is recommended that at least one type of disinfectant is used. Based on the findings of this study, a combination of chlorine and alum should be used as a household water treatment. This is due to the synergy effect which occurs from the combined method, where alum acts as a coagulant and chlorine as a disinfectant.

- The quality of the harvested rainwater in the rainy season is better than that harvested in the dry season. This is due to the longer period of dry antecedent days in the dry period, which causes more accumulation of sediment on the roof before rain collection. However, neither type of harvested rainwater met the World Health Organization's guideline limit for drinking water, thus necessitating the use of a household treatment technique.

- The study of the physical parameters in the tank revealed that they improved over time, notably at the top two levels of the tank, whereas the water quality at the bottom of the tank got poorer. The settling of the solids over time, with most of the settlement taking place in the first three days, was responsible for the changes in the parameters (particularly turbidity and TSS) in the tank (i.e., the improvement at the top and the deterioration at the bottom). The change in the microbial parameters in the tank also showed that the quality of rainwater at the top 2 levels improved with time while that at the bottom deteriorated over the tested periods. This was also attributed to the settling of bacteria that are attached to the solids. The investigated physical characteristics (i.e., turbidity, pH) and microbial parameters (i.e., total coliform and *E. coli*) did not meet the WHO recommendation limit for drinking water, despite improvements in these parameters at the top level over time. As a result, it is advised that at least one type of disinfectant should be used.

- The process of sedimentation not only reduced the solids but also decreased bacteria counts at the top of the tank. The enumerated bacteria increased at the lower depths of the tank. It is concluded that some of the bacteria are attached to the solids, and thus as the solids settle to the bottom of the tank, so does the bacteria.

- Alum, boiling, and chlorination are the three different household treatment methods used to treat rainwater collected from roofs. The results showed that while alum had a greater impact on lowering turbidity and particles in the water, boiling and chlorination had a greater impact on eliminating bacteria. It is thus recommended that a combination of chlorine and alum is applied to the stored water instead of individual treatment.

NOTES

1 The physical parameters of water, which include colour, taste, odour, temperature, turbidity, solids, and electrical conductivity, are determined by the senses of touch, sight, smell, and taste, while microbiological parameters include the presence of microorganisms such as helminths, bacteria, viruses, and protozoa (Davis and Cornwell, 2008).

2 A Pearson correlation is a number that ranges from −1 to +1 and describes the degree to which two variables are linearly related to one another.

Chapter 6

First Flush Impact on Rainwater Quality

ABSTRACT

Water scarcity is a huge problem in the under-developed and developing countries around the world, and consequently, rainwater can be an essential water source for meeting fundamental human needs. However, these nations have paid less attention to the efficiency of conventional rainwater treatments to improve the population's health. The impact of the first flush technique is particularly investigated in this chapter, and the microbial reduction resulting from its use was assessed from five different investigated roofs. Four roofs were constructed for the purpose of this study while the fifth roof was from an actual residential building. In this study, the microbial parameters (i.e., total coliform and *E. coli*) and the physical parameters (i.e., total suspended solids and turbidity) of the collected rainwater were studied. In every first flush experiment, a minimum reduction of around 40% from the initial pollutant load was observed. In addition, this study discovered that, as compared to other materials, corrugated galvanised roofing provided roof-harvested rainwater (RHRW) with the best microbiological quality. Although the first flush method has been evidenced here to increase the quality of RHRW, it is advised that household water treatment techniques (HHTTs) be implemented as well, especially if the collected rainwater is to be consumed.

6.1 INTRODUCTION

One of the many major problems faced by the economically under-developed nations of the world is the lack of clean drinking water. Studies have revealed that water scarcity in these nations has become a significant problem as the global population and water consumption have increased over the past few decades (Kaldellis and Kondili, 2007; Moruzzi and Nakada, 2015; WHO/UNICEF, 2019). According to numerical predictions and experimental studies, the impact of rising sediments and pollution levels in rivers with complicated topography and natural water resources can further destroy the already-existing sources of drinking water (Pu 2016, 2021; Pu et al., 2016, Zheng et al., 2016). The presence of contaminants (such as physicochemical contaminants

DOI: 10.1201/9781003392576-6

or microbiological pathogens) from the environment is the main concern about the usage of RHRW. Researchers have determined that the weather conditions when collecting rainwater and the impact of the age and cleanliness of the storage tanks, gutters, roof catchments, and pipes are the two main sources of external contamination (Yaziz et al., 1989; Lee et al., 2012; John et al., 2021a).

Water gradually accumulates in a roof's gutter system during a rainfall event before draining down through the downpipe. The raw rainwater from the roof, e.g., after a long period of dry antecedent days, can contain significant amounts of bacteria from decomposed insects and lizard, bird, and animal droppings as well as concentrated tannic acid. Additionally, it could contain chemicals, water-borne heavy metals, and sediments, all of which are undesirable additional pollutants to a water storage system. First flushing is a technique for reducing contamination in rainwater harvesting systems studied by Doyle (2006) and Martinson (2007). Individuals who harvest rainwater from different parts of the world practice first flush using numerous techniques. While some rainwater harvesters use first flush devices to divert the initial sets of rain, others drain out the rainwater collected at the initial period (John et al., 2021c). A first flush device diverts the initial sets of rainwater away from the storage tank, effectively washing the roof's dust, dirt, and other particles gathered from the environment over time. This initial flushing procedure typically contributes to improved water quality and the protection of water pumps and internal components of rainwater storage systems (Doyle 2006).

As a helpful intervention to reduce contamination in rainwater storage tanks, the first flush practice notion is becoming more widely recognised. The first flush method has certain advantages over the filtration method, reducing both dissolved and suspended contaminants, especially when trace minerals like zinc and lead are the problem (Poudyal et al., 2016; Paton and Haacke, 2021; Poudyal et al., 2021). The first flush procedure is also unaffected by particle size, which is particularly important when handling fine and lightweight roof dust. While the first flush is almost universally acknowledged to improve rainwater quality (Ntale and Moses, 2003; Doyle, 2006; Abbott et al., 2007), no consensus has been reached regarding the amount of rainwater that should be diverted or whether such diversion should be based on rain depth, roof size, rain intensity, volume, or rainfall duration (Doyle, 2006; John et al., 2021c).

Martinson and Thomas (2005) conducted a study to quantify the first flush phenomenon. While their results showed a wide variation, they suggested that the contaminated load will decrease by half for every millimetre of diverted first flush. Several studies have separately reached a similar conclusion (Doyle, 2006; Abbott et al., 2007; John et al., 2021c). The quality of rainwater from different roof materials has been a concern and has also been assessed by the literature (Yaziz et al., 1989; Uba and Aghogho 2000). In the Port Harcourt district of Rivers State, Nigeria, Uba and Aghogho's (2000) study examined the quality of rainwater from various roof catchments (i.e., corrugated galvanised iron sheet, aluminium, asbestos, and thatch roofing materials), and revealed that corrugated galvanised sheet provided the best harvested rainwater quality. The corrugated galvanised sheet's capacity for storing heat from high temperatures and ultraviolet light was mentioned as a factor that improved the rainwater quality.

6.2 EXPERIMENTAL METHODS AND AIM OF STUDY

6.2.1 Aim of Study

The goal of this chapter is to investigate and discuss the impact of first flush in improving the quality of RHRW. The physical and microbiological parameters of rainwater from four built roofs and from an asbestos roof on a residential building were used as benchmarks for the effectiveness of the first flush practice.

The objectives of the first flush study include:

• Determining the improvements of the first flush practice in enhancing the quality of roof-harvested rainwater.
• Observing the variations in the quality of roof-harvested rainwater from different types of roof types following the implementation of first flush.

6.2.2 Experimental Methods

The experimental approaches utilised to accomplish the objectives are presented in this section, whereby the measurement procedures were set up similarly to those utilised by Thomas and Martinson (2005) and Yaziz et al. (1989). Numerous studies have made recommendations regarding how much water should be flushed. Table 6.1 summarises some studies on the amount of rainfall needed for first flush analysis. Yaziz et al. (1989) recommended diverting 5 litres of the initial batch of rainwater during the first flush procedure. Diverting between 20 and 25 litres and 0.4 to 0.8 mm is suggested in other guidelines, such as the EnHealth Council (2004) and Texas Water Development Board (2005), respectively. A study conducted by John et al. (2021c) revealed that some households in rural areas of less economically developed countries sometimes ignore the practice of first flush.

The amount of rainwater used for the first flush analysis is 0.5 mm since one objective of this study is to examine the effects of first flush on both physical and microbiological qualities. The water sample was flushed six times in a row, each time after 0.5 mm of rain, for a total of 3 mm. This amount was selected because it falls between the minimum and maximum depth of rainfall that have been reported by various studies, as indicated in Table 6.1. On the other hand, sequential flushing presents the opportunity to monitor the improvement in rainwater quality over the successive 0.5 mm diversion. The annual rainy period in Nigeria runs from April to October. The measurements were conducted between May and July (although the analysed first rainfall event occurred only in June, but the roofs were constructed and completed in May).

Table 6.1 Literature review on first flush diversion (Doyle, 2006)

S/N	Reference	Specifications	Depth of the rainwater before first flush analysis
1	Pacey and Culis (1986)	Improve rainwater quality	First 10 minutes of rain event
2	Michaelides (1987)	Based on experimental work in Thailand	0.28 mm
3	Yaziz et al. (1989)	To protect tank against microbial contamination	0.33 mm
4	Cunliffe (1998)	Average size roof	20–25 litres
5	Ntale et al. (2003)	Empirical value	0.83 mm or first 10 minutes
6	Martinson and Thomas (2005)	Sample measurements	0.4–0.6 mm
7	Texas Water Development Board (2005)	Depends on season, trees, dry days, and debris	0.41–0.82 mm
8	Rainwater harvesting (undated)	Minimum, low, and high pollution, respectively	0.2 mm, 0.5 mm, and 2 mm, respectively

6.2.3 Experimental Methods from a Selected Roof in the Ikorodu Area

The measurements involved collecting first flush water from a chosen asbestos roof top in Ikorodu for analysis, which is representative of the region. A 1 litre pre-sterilised plastic bottle was used to collect the harvested rainwater sample from the PVC piping that is linked to the gutter from the asbestos roof of a 2 m^2 area at various times during the rainfall event (Table 6.2). The asbestos roof in the chosen household was constructed 15 years before the measurements were taken. To achieve a realistic accumulation of particulate matter, it was made sure that at least three dry antecedent days had occurred before harvesting rainwater, as a procedure indicated by Martinson and Thomas

Table 6.2 Sampling program

Time since cessation of rainfall event	Turbidity	Suspended solid	Coliform bacteria (E. coli and total coliform)
Raw water (RHRW)	X	X	X
FFRW	X	X	X
First 0.5 mm	X	X	X
Second 0.5 mm	X	X	X
Third 0.5 mm	X	X	X
Fourth 0.5 mm	X	X	X
Fifth 0.5 mm	X	X	X
Sixth 0.5 mm	X	X	X

(2005) and (Doyle 2006). To avoid external contamination of the results, sterilised water was used to wash the gutter, PVC pipe, and plastic bottles prior to the collection of RHRW. The obtained water samples were kept in an icebox and transported immediately to the laboratory for examination.

The precipitation of rain (in mm) was measured using a rain gauge. Before being set up to collect rainwater in an open area free from trees, houses, or other structures, the rain gauge was also cleaned with sterilised water. This process was repeated after each measurement of rainfall. To allow for the completion of six successive analyses of 0.5 mm of rainwater, a minimum of 4 mm of precipitation was recorded by the rain gauge for each studied rainfall event (Table 6.2). The samples were gathered during the continuous period outlined in Table 6.2. While the roof-collected rainwater was taken for each time and subjected to both physical and microbiological investigations, the free-fall rainwater was collected using a pre-washed plastic container that was placed in an open area, avoiding interference from any buildings, trees, or other structures.

A 255-litre storage tank (Figure 6.1) was used to collect the rain from the gutter. The analysed microbial parameters include total coliform and *E. coli*, while the analysed physical parameters measured include TSS and turbidity. These parameters were chosen because they have tendencies to change from

Figure 6.1 Experimental set up showing the first flush diverter

the start of the rain to the end. The flush division test outlined in Table 6.2 was performed for five different rainfall events to ensure a reasonable conclusion could be drawn. Furthermore, microbiological testing was carried out twice with separate samples to ensure consistency of results.

6.2.4 Experimental Methods for Different Roofs

Asbestos, aluminium, corrugated galvanised iron sheet, and plastic sheet were among the materials used to construct the different roofs, as shown by the examples in Figure 6.2. Different newly constructed roof materials were cleaned with sterile water before being used for the first flush analysis. These roofs were built 1.5 m above ground level. The same first flush analysis was used for the residential building in Section 6.3.1.

6.2.5 Colilert-18 Method

The *E. coli* and total coliform concentrations were enumerated using the standard Colilert-2000® procedure for analysis. The Colilert technique was executed in accordance with the manufacturer's guidelines, detailed in the standard methods in Chapter 3; numerous studies, including Julian et al. (2015); Martin and Gentry (2016), and John et al. (2021c), have used this approach. All the microbiological testing conducted for this study was done twice.

Figure 6.2 The aerial and side view of the four constructed roofing materials before applying on field

Table 6.3 Rain event characteristics

Rain event number	Total storm depth (mm)	Dry antecedent period (days)
R1	23.7	12
R2	15.4	5
R3	8.9	3
R4	11.6	6
R5	6.2	4

Notes: R1, R2, R3, R4, and R5 denote the first, second, third, fourth and fifth rain event, respectively.

6.2.6 TSS and Turbidity

The turbidity was examined using a Hanna Turbidimeter and the TSS were determined using a vacuum filtration device. The procedures utilised to determine TSS and turbidity in the water samples follow those outlined in the APHA, AWWA, and WEF standards (2005). Duplicate results of each test were also obtained.

6.3 RESULTS

Results from the first flush impact on rainwater from a single residential roof and four newly built roofs in the Ikorodu area of Lagos State are presented in Sections 6.3.1 and 6.3.2 of this chapter. Table 6.3 illustrates the characteristics of the harvested rainfall events.

6.3.1 Water Quality Analysis from a Single Residential Roof

This section examines how first flush influences the quality of rainwater collected from an aging asbestos roof on a representative residential building. The result is analysed in two parts, addressing the physical and microbial parameters. The discussion in this section analyses the results obtained from all the water quality tests to comprehend the impacts of diverting and flushing several millimetres of roof runoff.

Physical Parameters

The physical parameters analysed in this section include total suspended solids and turbidity. These physical parameters were chosen because they can change over a period. This section also presents an analysis of the percentage reduction in the analysed parameters for five separate rainfall events (Table 6.4). The percentage reduction was calculated by subtracting the successive millimetre and ratioing it with the initial value.

The outcome demonstrated that RHRW has higher turbidity and TSS values compared to harvested free-fall rainwater (FFRW). The accumulation of

Table 6.4 Percentage difference of the measured parameters for the five rainfall events

Measured parameters	Rain events	1 mm	2 mm	3 mm	Total reduction
TSS	R1	47.1	27.7	10.5	85.3
	R2	45.9	21.1	11	78
	R3	53.1	25.5	11.2	89.8
	R4	48	18.6	10.8	77.4
	R5	51.1	27	9.2	87.3
Turbidity	R1	49.5	26.7	8.8	85
	R2	55.2	24.1	12.6	91.9
	R3	52	25.1	10.4	87.5
	R4	50.8	27.8	9.4	88
	R5	48.1	24.5	10.6	83.2
Total coliform	R1	45.8	26.1	13	84.9
	R2	38.4	27.4	14.5	80.3
	R3	48.6	17.1	12.1	77.8
	R4	48.1	15.1	17.6	80.8
	R5	47.4	16.2	15.4	79
E. coli	R1	42.5	28.9	12.2	83.6
	R2	42.4	24.1	12.6	79.1
	R3	47.3	22.3	12.9	82.5
	R4	46.9	21.8	13.2	81.9
	R5	48.3	22.8	11.4	82.5

dust, particle matter, and animal and bird droppings on the roof, especially following a long period of dry antecedent days, was the cause of the elevated turbidity and TSS values. Such rooftop contaminants have been suggested by Lee et al. (2012) and John et al. (2021b) to not only contaminate RHRW but also lead to sediment build-up in the rainwater storage tank.

The results demonstrated that TSS and turbidity levels decreased steadily throughout the course of the six subsequent flushes (Figures 6.3 and 6.4), where each flush represents a 0.5 mm diversion that adds up to a total of 3 mm. For each of the five rainfall events, the decrement percentage in these parameters is shown in Table 6.4. After the 3 mm of rainwater was diverted, the examination of these results revealed a 77% reduction in TSS and 83% reduction in turbidity. The initial 1 mm diversion for TSS and turbidity showed the greatest reduction, while the third 1 mm diversion showed the least reduction.

The results revealed that 45% of the total reduction of 77% of TSS occurred after the first 1 mm diversion, and 48% of 83% for turbidity. A 10% reduction of TSS occurred in the last 3 mm diversion, while 9% was observed for turbidity (Table 6.4 and Figures 6.3 and 6.4). Although the turbidity in RHRW improved after the first flush procedure, the results showed that most turbidity levels (post the 3 mm diversion) were still above the WHO recommended limits of 5 NTU (presented by the red line in Figure 6.4). *FF* in Figures 6.3 and 6.4 denote first flush.

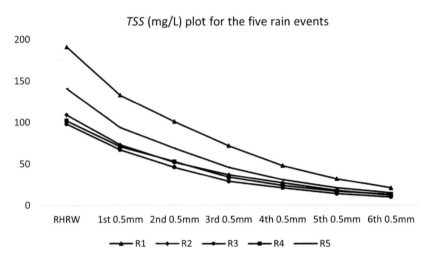

Figure 6.3 TSS values (mg/L) for the five rain events in increments of 0.5 mm

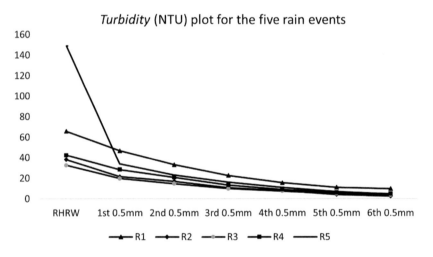

Figure 6.4 Turbidity values for the five rain events in increment of 0.5 mm

Microbiological Parameters

The results in this section present the microbes (i.e., total coliform and *E. coli*) over six successive periods for five rainfall events. This section also presents the percentage difference in each successive millimetre of diverted rainwater (Table 6.4). Each time 0.5 mm of rainwater was diverted, there was a reduction in the number of microorganisms. The outcomes demonstrated that the total coliform and *E. coli* of the collected FFRW were lower than those seen in RHRW (Figure 5.3). As previously discussed, the higher

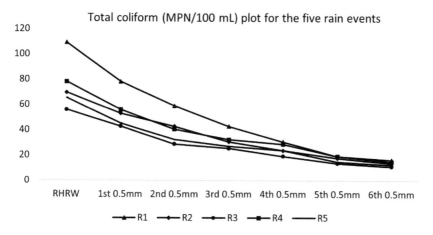

Figure 6.5 Total coliform values (MPN/100mL) for the five rain events

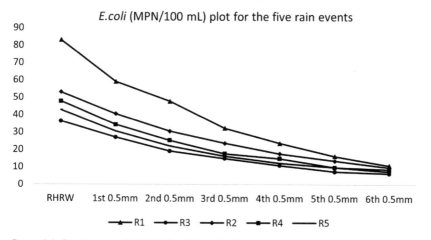

Figure 6.6 E. coli values (MPN/100 mL) for the five events

values were a result of dust/particulate matter and animal waste accumulation on the roof, especially following a long period of dry antecedent days.

Figures 6.5 and 6.6 portray a steady decrease in total coliform and *E. coli* across each succeeding flush for all five rainfall events. After the 3 mm of rainwater was diverted, the analysis of the results revealed about 78% reduction in the quantity of microbes. Most of the reduction was seen in the first millimetre diversion (i.e., 39% of the removed 78%), while subsequent analysis showed the least reduction happened in the third millimetre diversion (i.e., 12% of the removed 78%) (Table 6.4 and Figures 6.5 and 6.6). The results of this investigation demonstrate that the rainwater quality is still unfit for drinking as benchmarked by the WHO guideline, despite the first-flush procedure having the ability to reduce the contamination in RHRW. Thus, when RHRW is not disinfected, there is risk involved in drinking it. Although various studies have suggested diverting between 0.3 and 2 mm (Yaziz et al., 1989; Ntale and Moses 2003; Martinson and Thomas 2005; John et al., 2021b), the findings from this study based on the proposed pilot results collected from Ikorodu have indicated that higher levels of diversion may be necessary (John et al., 2021b).

6.3.2 Impact of First Flush on the Four Constructed Roofs

The outcome of this section's analysis of physical (i.e., TSS and turbidity) and microbiological characteristics (i.e., total coliform and *E. coli*) is to show the impact of first flush for various roof materials. The results from the presented measurements were studied for six successive diversions of 0.5 mm over five rainfall events.

Physical Parameters

To explore the physical parameters of the collected rainwater, Figures 6.7 and 6.8 present the plots of TSS and turbidity after six successive diversions of first flush (i.e., 0.5 mm each flush). Consistently for all rainfall events, the results showed that the physical quality of the RHRW is poorer than those of the FFRW (Figures 6.7 and 6.8). As mentioned in the preceding section, the build-up of dust and faecal depositions on the roof is one of the causes of these greater turbidity and TSS values.

The results also illustrated that TSS and turbidity readings continuously decreased throughout the course of six subsequent 0.5 mm diversions for all five rainfall events. It is also worthed noting that the majority of TSS and turbidity reduction took place during the first millimetre of flush. In addition, each millimetre of rainwater diversion resulted in an average of 40% reduction for TSS and turbidity. This result is agreeable with the study undertaken by Martinson and Thomas (2005), which suggests about a 50% reduction. The study by Martinson and Thomas (2005) was done in Uganda, Ethiopia, and Sri Lanka. The outcome of the physical parameters in this section further demonstrated results consistent with those in the previous section, where an existing residential roof was investigated. Corrugated galvanised sheet gave the best quality of rainwater while asbestos gave the worst quality. Despite the improvement of the quality of the RHRW, results showed that the rainwater quality failed to meet the World Health Organization's guideline for turbidity.

Microbiological Parameters

This section studies the changes in total coliform and *E. coli* over six separate intervals of 0.5 mm rainwater flushes (refer to Figures 6.9 and 6.10). Similar

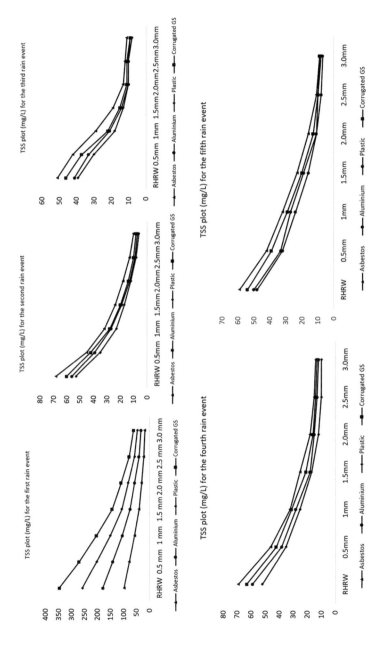

Figure 6.7 TSS plot (mg/L) for the five rain events

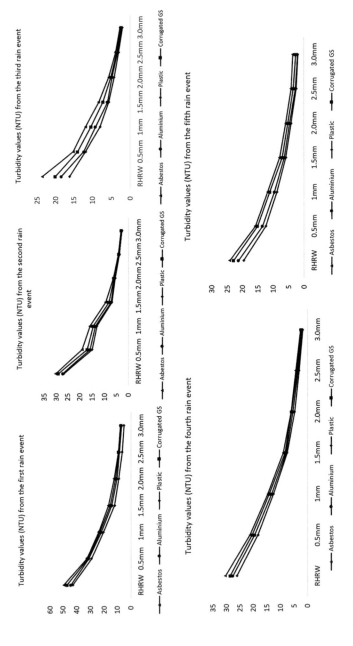

Figure 6.8 Turbidity plot (NTU) for the five rain events

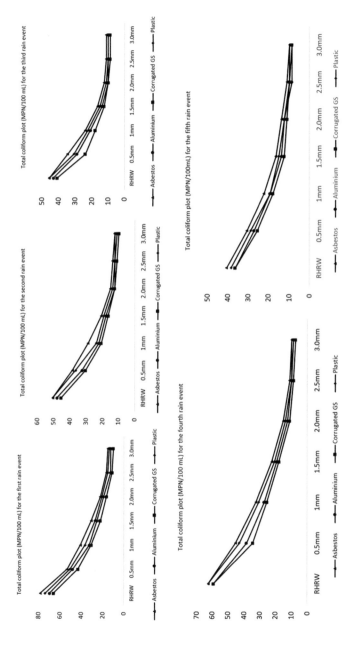

Figure 6.9 Total coliform plot (NTU) for the five rain events

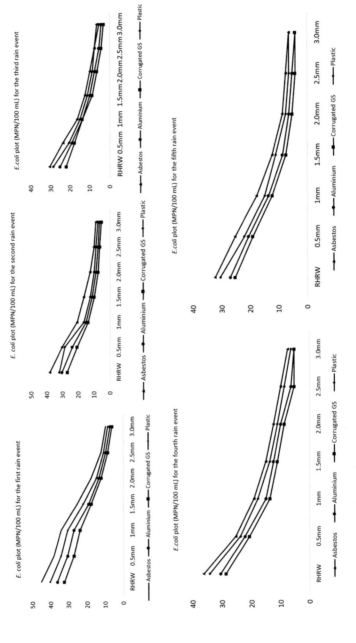

Figure 6.10 E. coli plot (NTU) for the five rain events

to what was discovered on the residential building roof, the statistics showed that the FFRW has a lower bacterial level than the RHRW. Additionally, the data also revealed that the bacterial level decreased steadily over the course of six successive diversions for each rainfall event, with the majority (i.e., over 39%) of the bacterial reduction taking place during the first millimetre of rainwater diversion, and at least a 10% reduction between each millimetre of flush (Table 6.4).

Asbestos roofing produced rainwater of the lowest quality compared to the other roof types in most events and flushes. Aluminium and galvanised corrugated sheet roofs can be impacted by ultraviolet radiation to produce high temperatures, which can facilitate disinfection of the captured rainwater, hence causing a comparatively greater quality of collected rainwater. Plastic roofing provided intermediate microbiological parameter results among all roof types due to its nature and facilitation of temperature capturing (Figures 6.9 and 6.10). On the other hand, the quality of RHRW collected from the new asbestos roof is superior to that gathered from the 15-year-old asbestos roof (physical view for both at Figure 6.11). Their discrepancy findings revealed at least a 15.9% decline in the physical and microbiological parameters during the same rainfall event (Table 6.5). This suggests that one of the sources of contamination for RHRW was the wear-off residual material from the used asbestos roof.

The measurements in this study also evidenced that both the physical and microbiological parameters (such as TSS, turbidity, total coliform, and *E. coli*) were reduced by at least 75% over the course of the 3 mm diversion for each

Figure 6.11 Asbestos roof material from the newly constructed and the household asbestos roof material, respectively

Table 6.5 Percentage difference obtained on the results from both old and new asbestos

Rain events	TSS (mg/L)			Turbidity (NTU)			Total coliform (MPN/100 mL)			E. coli (MPN/100 mL)		
	Old asbestos	New asbestos	% Diff	Old asbestos	New asbestos	% Diff	Old asbestos	New asbestos	% Diff	Old asbestos	New asbestos	% Diff
R1	191	98	48.69	65.8	49.5	24.77	109.1	78.2	28.32	83.1	45.3	45.49
R2	109	68	37.61	38.2	30.3	20.68	69.7	50.4	27.69	53.1	38.4	27.68
R3	98	51	47.96	32.7	23.6	27.83	56	45.3	19.11	36.4	30.6	15.93
R4	102	69	32.35	42.5	30.4	28.47	78.2	62.4	20.20	47.8	36.4	23.85
R5	141	59	58.16	148.9	24.2	83.75	65.4	40.6	37.92	42.9	32.4	24.48

Notes: R1, R2, R3, R4, and R5 denote the first, second, third, fourth and fifth rain event, respectively.
% Diff. denotes percentage difference

rainfall event. According to Martinson and Thomas (2005), first flush can effectively remove 85% of the contaminant load. However, both their study and the one under discussion here consistently showed that the first flush procedure will not totally clean the rainwater of contaminants. Thus, it is advisable to employ household water treatment techniques, particularly if the harvested rainwater is to be used for human consumption. The outcomes from the field data in this study also revealed that the quality of the harvested rainwater following a longer period of dry antecedent days was poor due to the deposition of more atmospheric contaminants on the roof. This conclusion has been reached after considering the analysis of Figures 6.3–6.10 and Tables 6.4–6.5.

6.4 DISCUSSION

The concept of first flush has been explored in the literature and is practised in different forms. For instance, some rainwater harvesters use first flush devices to divert the initial rain, while others prefer to ignore the first set of rain or practice non-collection over an initial rain period. John et al. (2021c) showed that some residents wait for minutes, while some do not harvest the first rainfall event for 3–4 days. Also, many reservoirs in the UK feature a bypass canal at the reservoir's head to direct the first flush of rainwater around its perimeter. In short, first flush is a technique practised by many globally (Doyle, 2006; John et al., 2021b) that has merits and demerits. First flush helps to reduce the overall contamination of RHRW, thereby reducing the cost of HHTTs. Furthermore, using the first flush technique lengthens the life of pumps and other harvesting equipment, as well as lowering the need for tank maintenance. However, the volume of the harvested rainwater will be reduced after flushes, and this will be disadvantageous especially in areas where rainfall is rare.

The quantity of rainwater to be diverted during first flush has been a subject of discussion around the world. A study carried out from March to April 1980 on RHRW from an office building in central Tokyo concluded that between 0.5 and 1 cm of rainwater were required to wash the roof, while 1.5–2 cm can make the water quality improve steadily (People for Rainwater Japan, 2003; Freeflush, 2017). Doyle et al. (2006) further recommend that 1 mm diversion of rainwater would improve the quality of harvested rainwater. Many other studies, including those listed in Table 6.1, have given different ranges of values that should be diverted.

The quantity of rainwater to be diverted in a first flush system should not only depend on the environment's contaminant characteristics and roofing materials, but also on the purpose of the rainwater (Doyle, 2006). Although the probability that the first flush practice will produce entirely clean water is very low, the research outcome from this study has proven that the quality of harvested rainwater can be improved after first flush. Regardless of the first flush practice, there is some risk associated with drinking roof-harvested rainwater unless it has been disinfected or chemically treated.

Although several literature recommended diverting between 0.3 and 2 mm (Pacey and Culis, 1986; Michaelides, 1987; Yaziz et al., 1989; Cunliffe, 1998; Ntale et al., 2003; Martinson and Thomas, 2005; Texas Water Development Board, 2005), a recommendation of at least 2 mm of runoff diversion is suggested from the findings of this study before harvesting the rainwater into the storage tanks (based on the results obtained from Ikorodu district of Lagos State, Nigeria). The need to correctly determine first flush diverting range is required for optimal performance, and this is affected by many variables, such as gutter materials, slope, and size; air quality; rain frequency, duration and intensity; period of dry antecedent days; and wind speed and direction.

The chance of filling the storage tanks is low when first flush is significantly increased, especially in areas of low rain intensity and water scarcity, while the tank will be filled with sediments and contaminants if the first flush is too insignificant (John et al., 2021c; Savou, 2021). It is imperative that first flush diverters and other materials used in harvesting the rainwater are properly and regularly cleaned to serve their purpose. First flush diverters should be made to have a continuous drain for emptying them between rains, but this can have drawbacks due to the amount of water loss that can occur (Savou, 2021).

Another imperative factor that affects the quality of the RHRW in first flush practice is the type of roofing material. Results from this study's analysis of the microbial parameters from the various roof types revealed that the asbestos roofing material produced the harvested rainwater with the lowest quality, while aluminium and galvanised corrugated sheet, and arguably plastic roof types, harvested a relatively higher quality of rainwater (Figures 6.7–6.10). The ability of these roof types to absorb ultraviolet radiation and high temperatures, which quickly disinfect the harvested rainwater, can improve the resulting microbiological quality of the roof-harvested rainwater. Due to its greater capacity to absorb and store heat compared to other roof materials, corrugated galvanised roof produces the rainwater with the highest quality among the tested materials (Martinson and Thomas, 2005; Doyle, 2006; Lee et al., 2012; John et al., 2021b).

Corrugated galvanised sheet has consistently produced the best quality of harvested rainwater, according to a study that evaluated the quality of rainwater collected from four different roofing materials (i.e., also including wooden shingle tiles, clay tiles, and concrete tiles) (Martinson and Thomas, 2005; Lee et al., 2012). Martinson and Thomas (2005) conducted a study to quantify the first flush phenomenon on four different roofing materials (i.e., corrugated galvanised iron sheet, corrugated asbestos sheet, clay tiles, and tar sheet), and their findings revealed that the corrugated galvanised iron sheet and the tar sheet provided the best and worst quality of harvested rainwater, respectively. However, analysis of the data revealed that the plastic roofing material provided the highest level of quality for the physical characteristics examined (i.e., TSS and turbidity). In a study by Uba and Aghogho (2000) in which they examined the quality of rainwater from various roof catchments in the Port Harcourt district of Rivers State, Nigeria (i.e., catchments with corrugated galvanised iron sheet, aluminium, asbestos, and thatch roofing materials), their results correspondingly evidenced that corrugated galvanised sheet harvesting leads to the best rainwater quality.

The age of the roof also plays a significant role in the quality of the RHRW. The data from both asbestos roofing materials (i.e., from the existing household in Section 6.3.1 and the newly built roof in Section 6.3.2) reveals that the recently constructed roof collects rainwater with better quality. The comparisons of the residential building and newly built asbestos roof indicated that the latter causes at least a 60% reduction in the physical and microbiological parameters during the same rainfall events. The older asbestos roof's tendency to easily wear off, contaminating the collected rainwater, explains this discrepancy.

The results from the presented field measurements also demonstrate that longer periods of dry antecedent days lead to a lower quality of roof-harvested rainwater when compared to shorter dry antecedent periods. Contamination from bird droppings, dust, and other airborne contaminants that build up during long dry antecedent periods is a contributory factor to a lower quality of captured rainwater. Tables 6.4 and 6.5 present the quality of the harvested rain events, and Table 6.3 displays the characteristics of the rainfall events that include periods of dry preceding days. The correlative results from these tables demonstrate that as the number of dry antecedent days increases, the quality of the harvested rainwater deteriorates.

The experiment in Section 6.3.1 reveals a minimum of 78% reduction in the physical and microbiological parameters (i.e., TSS, turbidity, total coliform and *E. coli*) across the six consecutive rain diversions to a total of 3 mm, which is consistent for the five investigated rainfall events (Table 6.4). According to the study by Martinson and Thomas (2005) to assess the effect of first flush on water quality, 85% of the entering contaminant load can be eliminated using an effective first flush technique. However, Hathaway et al. (2011) evaluated the effect of first flush on indicator bacteria and TSS in stormwater runoff and suggested that the initial flush has stronger influence on TSS than the indicator bacterium, which included *E. coli* and enterococci.

Conclusively, the results within this chapter proves that first flush has a positive effect on harvested rainwater. However, it is clear that first flush does not totally remove contaminants from the water. Irrespective of the amount of rainwater diverted, the quality of the rainwater was found to be unsuitable for potable use because the examined physical and microbial parameters did not meet the World Health Organization's guidelines for drinking water. Therefore, it is recommended that the effective application of a form of HHTT is needed to disinfect the rainwater before use or consumption.

6.5 CONCLUSION

This chapter examined the extent to which the first flush technique improves the quality of RHRW from a sampled residential roof and four different newly built roofs in the Ikorodu district of Lagos State, Nigeria. It is advised that at least 2 mm of the rain runoff is diverted. The measurement outcomes indicate that every 1 mm of the pollutant load saw a minimum reduction of approximately 40% from the initial contamination level. Furthermore, it was discovered that corrugated galvanised roofing material produced better results in terms of the microbiological quality of roof-harvested rainwater, while plastic roofing material produced better results in the physical quality of roof-harvested rainwater. Finally, even though the first flush technique has been proven to increase the quality of roof-harvested rainwater, it is essential to use a form of effective HHTT, especially if the harvested rainwater is intended for consumption.

Chapter 7

Health Risk Assessment for Potable Use of Rainwater

ABSTRACT

The Quantitative Microbial Risk Assessment (QMRA) technique is used to enhance water safety planning and evaluate microbial risks, in which anticipated health implications are explicitly quantified. Roof-harvested rainwater (RHRW) in the Ikorodu neighbourhood of Lagos State, Nigeria, is evaluated in this chapter, and recommendations to reduce adverse health risk to the residents are proposed. In the study region, various types, designs, and uses of rainwater harvesting systems have been assessed to determine the risk of human exposure to Escherichia coli (*E. coli*). To accomplish these objectives, a thorough survey of 125 households was conducted, and outcomes revealed that 25% of respondents consume RHRW. The risk of exposure to hazardous *E. coli* from RHRW utilised as potable water has been quantified using QMRA methodology, based on the use of 2 L of rainwater per day per capita. The highest *E. coli* exposure risk from consuming RHRW, both without the use of any household water treatment techniques (HHTTs) and with the use of alum treatment, was determined to be 100 and 96, respectively, for the projected number of infection risks per 10,000 exposed households per year. This estimated QMRA was obtained based on the premise that 7% of *E. coli* is viable and harmful. To reduce the risk of exposure to harmful *E. coli*, RHRW must be treated with an appropriate disinfectant prior to use.

7.1 INTRODUCTION

Up from 70% in 2015, 74% of the world's population has access to services for securely managing drinking water in 2020. In 2020, however, two billion people still lacked access to secure drinking water resources, including 1.2 billion who do not have the most basic level of service. It was further estimated that 1.6 billion people will not have access to properly managed drinking water by 2030 if current rates of progress are not accelerated. Rural areas account for eight out of every ten individuals who lack access to the most basic drinking water services, with least developed nations accounting for nearly half of them (UN, 2020). In less economically developed nations,

DOI: 10.1201/9781003392576-7

access to clean, drinkable water is not a norm, and most individuals rely on solitary gathering methods such wells, rainfall collection systems, boreholes, and rivers (Balogun et al., 2017; John et al., 2021d). Thus, researching the effects of rainwater harvesting is essential, especially in rural areas where access to community pipe-borne clean water is scarce.

Estimating the health risk associated with drinking water is crucial due to its influence on human well-being and longevity, as was covered extensively in Chapter 2. Studies show that the Quantitative Microbial Risk Assessment (QMRA) and epidemiological methodologies are two of the most prevalent ways to evaluate microbial risk in drinking water (Calderon et al., 1991; Haile et al., 1999; Colford et al., 2012). While the epidemiological method informs the propensity of these elements implicitly, the QMRA specifically identifies the origin of faecal pollution, the fate and kinetics of the microorganisms, the natural variability of the bacteria in the environmental matrix, and the etiological agent (Whelan et al., 2014). The QMRA procedure can be especially tailored to quantify pathogen exposure in humans (John et al., 2021d). Furthermore, numerous studies have demonstrated that, in situations where epidemiological research may be impracticable, QMRA provides accurate interpretation of the epidemiological results by estimating the risk to human health (Pruss, 2002; Zmirou et al., 2003; Whelan et al., 2014).

In other studies, the use of QMRA has been examined to evaluate the potential health risk associated with (1) ingestion of water and other uses including swimming and recreation; (2) primary sewage; (3) human enteric viruses; and (4) swine, dog, cattle, fresh gull, sea gull, and cattle faeces (Calderon et al., 1991; Soller et al., 2006; Roser et al., 2006; Soller et al., 2006; Ahmed et al., 2010; Ashbolt et al., 2010; Schoen and Soller et al., 2010; Whelan et al., 2014, John et al., 2021d). These studies have consistently used four sets of information to estimate the potential risks: (1) hazard identification, (2) dosage response, (3) exposure assessment, and (4) risk characterisation and management (Haas et al., 1999; Hunter et al., 2003; Ahmed et al., 2010; Whelan et al., 2014). Additionally, studies have been carried out to investigate potential microbial risks of harvested rainwater based on the presence of bacteria in rainwater storage tanks (Lye, 2002; 2009; Schets et al., 2010; Soller et al., 2010; Vialle et al., 2012; Lim and Jiang, 2013; Whelan et al., 2014; John et al., 2021d). These studies have explored pathogenic strains of *Legionella pneumophila, Salmonella species (spp.), Clostridium perfringens, Enterococci spp., Camplyobacter spp., Cryptosporidium spp., Giardia spp.*, and *E. coli* to assess the risk posed by microbes to human health. Rainwater storage tanks examined in various locations in the world (such as USA, Holland, Denmark, Greece, Uganda, France, Nigeria, and Australia) contained these pathogens. Multiple bacterial techniques, including Colilert, qPCR, PCR, and membrane filtration, were used by these researchers on their respective samples of rainwater (Gerba et al., 1996; Ahmed et al., 2010; Lim and Jiang, 2013; Machdar et al., 2013; Whelan et al., 2014; John et al., 2021d). This implies that a variety of pathogen enumeration techniques and water samples with a wide range of characteristics can be used to perform QMRA, and shows its versatility.

Dependence on rainwater is high in Lagos and especially in Ikorodu (Longe et al., 2010; Balogun et al., 2017), and with its prevalent lack of access to clean water, the health risks associated with drinking RHRW must be evaluated. Therefore, the purpose of this chapter was to evaluate the risk of ingesting roof-harvested rainwater in the Ikorodu area of Lagos State using QMRA. This chapter details the risks and best practices for the safe and sustainable use of rainwater. The development of the QMRA in this case study has also followed suggestions from other studies (George et al., 2015; Soller et al., 2010; Abia et al., 2016), and the used pathogenic strain is *E. coli* as it has been proven to be the most severe strain in this region (John et al., 2021c,d).

Various scenarios discovered through the fieldwork survey have been used to assess the hazards associated with *E. coli* exposure within this chapter. In total, 125 carefully selected representative households had consented to participate in the survey, in which the structured questionnaires were administered to them for completion. To evaluate the population's health in relation to those who consumed untreated harvested rainwater, the survey data and number of bacteria were co-applied in a compensated manner, broadening the analysis to include the entire study area.

7.2 METHODS

7.2.1 Study Area and Administration of Questionnaire

The Ikorodu Local Government Area of Lagos, Nigeria (refer to Figure 7.1) is located in the northern part of Lagos, approximately at latitude 6° 36' north of the equator and longitude 3° 30' east of the Greenwich meridian (Maplandia, 2015). The residents of that area are dependent on rainfall, especially during the rainy season, and it is also one of Lagos's fastest-growing neighbourhoods (Longe et al., 2010; Ukabiala et al., 2010). The collection of rainwater has an impact on the diverse communities because this location comprises a variety of business and retail establishments, residential buildings, and governmental and private organisations. Due to this feature, the study is both challenging and significantly important for the subject matter.

The questionnaires were distributed to households located in a mix of good and poor sanitation areas using a house-to-house survey technique. This survey method was selected due to the availability and ease of access to each household in the studied Ikorodu area, and the difficulty of obtaining detailed information via the internet (due to internet coverage). Also, when administered, the survey and its targets can be confirmed, as compared to an internet survey that might risk being filled by non-residents. The sampling error was calculated to be approximately 0.5% using a 95% confidence interval. The map and population density of the study area were analysed, and a thorough inspection was conducted before classifying the regions. To

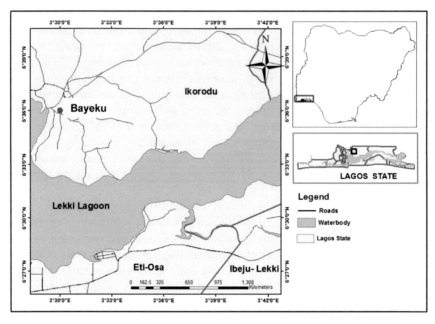

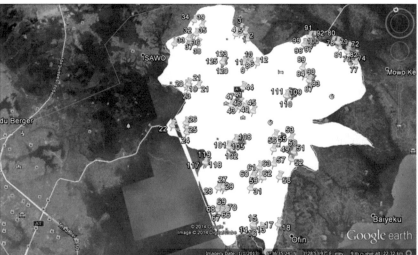

Figure 7.1 Map of Lagos State in cross-section, highlighting Ikorodu and the locations where questionnaires were distributed (Google Earth, 2020; Umunnakwe et al., 2019; John et al., 2021d)

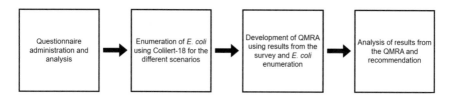

Figure 7.2 Flowchart of the methodology

ensure correct responses from the respondents, the questionnaires were given to a person between the ages of 25 and 50 in each of the households.

The authors provided support by explaining the surveys. Figure 7.1 illustrates the distribution of the participating households. There is an average of 4.9 people in each household for the 125 surveys that were collected. Only 1% of the total of 613 persons were above the age of 60, as compared to 11% of children under the age of 5. This information is crucial considering germs may affect these groups more severely if ingested. The QMRA was developed using these demographic findings as well as the quantity of pathogenic *E. coli* bacteria consumed from the rainwater storage tank. All participating households provided completed questionnaires. Results from the survey were examined in two stages: (1) water usage strategy and (2) water and sanitation infrastructure. The questionnaire used is located in Appendix E.

A simplified flow chart for the research process is presented in Figure 7.2. Firstly, the surveys were distributed to different areas of Ikorodu. To estimate the level of *E. coli* exposure to the population, samples were collected according to the Colilert-18 method that is being used to conduct bacterial counts for the RHRW. The proposed QMRA analysis was then developed using the data gathered from the examination of questionnaires and the enumerated *E. coli*. The experiments for the scenarios involving the use of different HHTTs used by residents of the Ikorodu neighbourhood of Lagos are described in Section 5.2.2. The HHTTs used are alum, chlorination, boiling, and the combination of alum and chlorine. The results from that section were applied to the QMRA model. Finally, the examination of the QMRA was conducted, and recommendations were made.

7.2.2 Development of the QMRA

The previously outlined four-phase process is used during the QMRA to estimate the risk to human health posed by exposure to the target microorganisms. Gerba et al. (1996), Ahmed et al. (2010), and John et al. (2021d) provide detailed descriptions of this process comprising: (i) hazard identification; (ii) exposure evaluation; (iii) dose–response evaluation; and (iv) risk characterisation. These phases are explained as follows.

Hazard Identification

This phase is completed by compiling data on the presence of the targeted pathogen using different household water treatment techniques (HHTTs). Using the established Colilert-18 procedure, the target pathogen's presence (i.e., positive/negative) and its number are counted in both the storage rainwater tanks and roof (i.e., APHA protocol number 9223 B. Enzyme substrate test). This phase represents an initial assessment of data and is scrutinised more meticulously in the next phases of the QMRA process to facilitate the identification and categorisation of all the risks involved in consuming rainwater. This phase facilitates the decision on whether there is sufficient information to consider a substance (e.g., *E. coli*) as the cause of adverse health, including diarrhoea (Lammerding and Fazil, 2000; Ahmed et al., 2010). The following sections describe the target pathogens and the sampling methods used.

Studied Pathogen

E. coli was this study's target pathogen because of its key role in causing adverse effect to human health. Given that numerous studies have described their prevalence in faeces, blood, food, urine, hands, water, wound swabs, etc., it poses a severe threat in the research region of Ikorodu (Akinyemi et al., 1998; Adejuwon and Mbuk, 2011; Olaleye, 2012; Johnson et al., 2014; John et al., 2021d, Sowunmi et al., 2022). There are hundreds of different strains of the bacteria *E. coli*, which is typically found in both people and animals' bodies (Weaver et al., 2015). Some strains, including *E. coli* O157:H7, have been identified to produce toxic by-products that can make humans ill. The dangerous *E. coli* strains were estimated to represent a maximum 7% of the whole *E. coli* population. (Ahmed et al., 2010). This estimation of 7% of harmful *E. coli* strains has been utilised in this investigation to evaluate the QMRA of the RHRW, and it has also agreed with the 8% suggested by Machdar et al. (2013). *E. coli* O157:H7 patients have a 30% increased risk of renal failure or high blood pressure (Kanarat, 2004; Lim and Jiang, 2013). Along with the possibility of a stroke or a seizure, these individuals may also experience acute stomach cramps or intestinal necrosis (tissue death). In some instances, the patient may have bloody diarrhoea, with symptoms setting in anywhere between a few hours and 10 days after the infection. Some who are sick can also transmit the disease to others even when they do not exhibit any symptoms (Kanarat, 2004; Lim and Jiang, 2013).

Sampling Methods

Colilert-18 method was used to enumerate the bacterial counts for all the investigated 49 rainfall events in both rainy and dry seasons. The details of all observed rain events can be found in Appendix G. This sampling exercise was used as it was proven by the manufacturer's analysis to be reliant with the ability to provide good statistical confidence in terms of sample size. The utilised Colilert-18 tests were executed in accordance with the manufacturer's guideline, where details of the Colilert-18 techniques have been described in Chapter 3.

Exposure Assessment

In this second QMRA phase, the quantity consumed by a person and the number of pathogenic *E. coli* bacteria in the rainwater storage tank were estimated. To calculate the chance of infection, this pathogen number was added to the dose–response models. The exposure assessment also evaluated the number of persons exposed, the categories of people impacted, and the degree and duration of exposure by each pathway (Petterson et al., 2007; Whelan et al., 2014). The pathway considered in this study included the microorganisms that were counted both before and after the application of various HHTTs.

Dose–Response Assessment

During this phase, the risk of a response was evaluated by a given dose (number of target pathogen microorganisms). The dose–response models, which are formed in statistical equations, specify how the target pathogen, transmission pathways, and hosts are related in the dose–response relationship. Utilising data sets from numerous human and animal investigations, the statistical model for predicting dose–response was suggested (Petterson et al., 2007; Whelan et al., 2014). The amount of pathogenic *E. coli* strains consumed before and after the usage of various HHTTs was measured in this study, and it was used to calculate the population's likelihood of contracting an infection.

Gerba et al. (1996), Ahmed et al. (2010), and Whelan et al. (2014) suggested using either an exponential or a Beta-Poisson model to evaluate the dose–response relationship. Strachan et al. (2005) further proposed that bacteria and viruses were better suited to the Beta-Poisson model compared to the exponential model (i.e., for protozoa and viruses). Thus, the following Beta-Poisson model was proposed since the pathogenic strains of *E. coli* were used as the study's target pathogen.

$$P(i) = u\left(1 - \left[1 + \left(\frac{d}{\beta}\right)\right]^{-\alpha}\right) \tag{7.1}$$

$$N = 10,000P(i) \tag{7.2}$$

In Equation 7.1, $P(i)$ represents the probability of infection per 10,000 people in the exposed population for a single event, while d denotes the dose (i.e., number of infective units). The percentage of harmful *E. coli* strains is denoted by the symbol u in Equation 7.1 (where 7% was used in this study as outlined and explained before). The number of illnesses per 10,000 people for a single incident is denoted by the symbol N in Equation 7.2. The best-fitting parameters for the Beta-Poisson model, α and β, are proposed to be 0.3126 and 2884, respectively, where their values were discovered through field observations and experimentation (Haas et al., 1999; Ahmed et al., 2010: Lim and Jiang, 2013). The pathogens in the water are assumed to be distributed randomly by the Beta-Poisson model. This model presupposes that variations in human reaction and pathogen competence cause the probability of infection per consumed pathogen to fluctuate among the exposed population (Haas et al., 1999). This presumption is true if $\alpha \ll \beta$ and $\beta \gg 1$.

According to Haas et al. (1999) and Leite and Moruzzi (2016), an accurate dose–response model for risk analysis should have a wide range of characteristics, including (1) a good imitation of human pathophysiology; (2) the capacity to analyse the exposure path in a manner comparable to that of a real infection; (3) a preference for infection as a response over adverse symptoms or death; (4) the possession of a pathogen strain that is similar to the infection's origin; (5) having statistically acceptable adjustment (do not reject null hypothesis, p > 0.05, for 95% significance); 6) modelling using information from two or more experiments and sets of data that are statistically similar; and 7) having a low average infectious dose quotient per average lethal dose (DI50 / LD50) in order to obtain a conservative risk estimate. In conclusion, it is challenging to find a single model with all above attributes, hence it is critical for each user to use the model that best suits their needs, as advised by Leite and Moruzzi (2016).

Data monitoring for a QMRA can be derived from indicator bacteria, distribution fit, and direct measurement of the pathogens (Ahmed et al., 2010). Direct counting of *E. coli* in the water samples produced the information used in this investigation. The administered well-structured surveys provided the *E. coli* exposure and risk scenarios.

Risk Characterisation and Management

Instead of providing a single number as for the first three phases, a range of values was provided by the Monte Carlo analysis in this phase, so that the results showed the whole range of potential risks from average to worst-case scenarios. This fourth step was used to combine data from the doses received (from the exposure assessment) and the risks connected to various doses (from the dose–response assessment), to determine the likelihood of risk. In short, data from the first three phases of hazard identification, dose response, and exposure assessment were combined into a single statistical model to estimate risk. This statistical model can be used to determine the likelihood that an infection, disease, or death will occur.

$$P_N = 1 - \left[1 - P(i)\right]^E \qquad (7.3)$$

$$N = 10000 P_N \qquad (7.4)$$

Equations 7.3 and 7.4 were used to calculate the probability and number of risk infections per 10,000 persons, respectively, in the Ikorodu neighbourhood of Lagos each year (Haas et al., 1999). In Equation 7.3, E stands for the annual number of exposure events, and P_N is the probability of infection per 10,000 rural Ikorodu people for various scenarios. In Ikorodu, the average number of days in the rainy season, which lasts from April to October every year, was used to calculate the value of E, which was found to be 281.

Risk management is part of QMRA. The goal of risk management is to reduce or eliminate risks and the bad consequences they could cause. Different approaches can be used for risk management, and it is most successful when influenced by risk characterisation (Petterson et al., 2007; Whelan et al., 2014). For the investigated case, risk management, including examining the effects and application of several HHTTs, has been deployed to suggest the best course of treatment.

E. coli as Target Pathogen: Advantages and Limitations

The detection of indicator bacteria like coliforms and E. coli is crucial for determining the microbiological quality of drinking water. As suggested by Ahmed et al. (2019), the faeces coliform group including E. coli is a more accurate indication of faecal contamination than other faecal coliforms. There are numerous strains of E. coli, and most of them are not harmful. However, some E. coli strains, like those that produce the Shiga toxin, can result in life-threatening sickness (WHO, 2016). However, it is also understood that the presence of pathogens does not always correspond with E. coli concentrations.

The study by Ahmed et al. (2010) revealed that the proportion of dangerous E. coli strains can be 7%, where similar conclusions on the harmful E. coli proportion have been established by Machdar et al (2013). Due to its viability, the capacity to reflect the potential presence of pathogens, and the simplicity and commonality of analysis, E. coli has been chosen as the microorganism to monitor in this research (WHO, 2016). The fact that E. coli does not always indicate the presence of pathogens in ambient waters is a significant drawback of employing it. Despite this, E. coli strains have been utilised as the primary source of the disease burden in numerous studies (Machdar et al., 2013; John et al., 2021c,d), and its analysis method has been widely employed to determine the microbial contamination of drinking water (George et al., 2015; Petterson et al., 2016; WHO, 2016; Ahmed et al., 2019). Furthermore, *E. coli* may be a strong indicator of lax sanitation practices, which are typical in low-income nations and where *E. coli* poses a major risk of contamination for children and older generations (WHO, 2016; Ahmed et al., 2019).

In hindsight, to evaluate rainwater treatment systems with respect to the risk to human health from faecal pathogens, generic faecal indicator bacteria such as *E. coli* are usually used as the screening tool before testing other pathogens (Petterson et al., 2016). Additionally, the ratio of these indicator bacteria to pathogens is occasionally employed for risk assessment (Petterson et al., 2016). In general, the presence of *E. coli* rather than its absence also serves as a stronger predictor of the possible existence of enteric pathogens (WHO, 2016).

7.3 RESULTS ANALYSIS

7.3.1 Analysis of the Questionnaire

The data from conducted survey was broken down into two categories for analysis: water use strategy, and water and sanitation infrastructure. The sections that follow explain their analyses.

Water Use Strategy

The results from the analysis of the questionnaires showed that boreholes are the main source of drinking water during the year. In contrast, the percentage of respondents who use boreholes decreased from 82% in the dry season to 55% in the rainy season (see Figure 7.3), emphasising the significance of rainwater usage among residents. According to research by Longe et al. (2010), 40 million people in Nigeria, or around 20% of the country's total population,

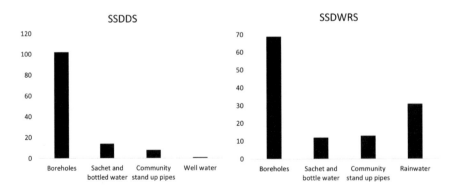

Figure 7.3 Sources of water for consumption throughout the dry and wet seasons, respectively (SSDDS and SSDWRS stand for sources of drinking water during the dry season and the rainy season, respectively)

Figure 7.4 Number of households that drink rainwater per household water treatment techniques (HHTTs)

harvest rainwater during the rainy season. In contrast to Nigeria as a whole, the findings from the Ikorodu region demonstrate that the dependence on rainwater is significant, with 114 households (out of the 125) harvesting rainwater for use. Among those 114 households, 31 of them drink rainwater (Figure 7.4). Furthermore, 6 of those 31 households acknowledged they drank untreated rainwater (Figure 7.4). The remaining 83 households collect rainwater for use in farming, washing dishes and clothes, taking showers, and other domestic needs. Despite the reliance on rainwater, the harvested rainwater is usually stored for a relatively brief period (between 2 days and 1 week, depending on the size of the family). To calculate the health risk for the various categories of rainwater drinker in the population, the information regarding domestic rainwater treatment was used in conjunction with the QMRA.

Water and Sanitation Infrastructure

Figure 7.5 demonstrates storage tank sizes for residents. In households with smaller rainwater tanks (20–225 L tanks), 77 households (roughly 76%), believed that there would be more contamination if the rainwater was stored for a longer period. The 24 remaining households (or roughly 24%) that collected rainwater were not concerned about how long it should be stored. Additionally, 25 of the 31 households who drank the gathered rainwater (i.e., 81%) reported being aware of the contamination in the water and had applied one form of HHTT, whereas the remaining six households (19%) had not applied any sort of treatment (also refer to Figure 7.4).

Figure 7.6 shows roof types for the community and the people who drank rainwater, while Figure 7.7 shows the rainwater harvesting methods used by all the households who collected and drank rainwater. The findings revealed that 83% of the study population collected rainwater into collection vessels before transferring it to storage tanks, while the remaining 17% gathered rainwater

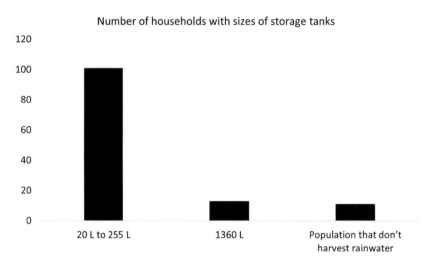

Figure 7.5 Sizes of storage tanks

directly into storage tanks. Further inspection evidenced that 88% of all roofs were corroded, where 84% belonged to surveyed rainwater harvesters. It is crucial to know this information since previous research has indicated that old roof materials contribute to RHRW contamination (Uba and Aghogho, 2000).

Figure 7.8 presents the statistics of the first flush procedure used by rainwater harvesters. As suggested in numerous studies, first flush has increased the quality of RHRW (Kus et al., 2010; Amin et al., 2013; John et al., 2021d).

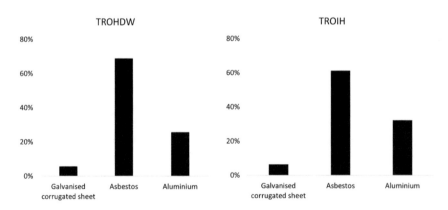

Figure 7.6 Type of roofs used by the residents of Ikorodu (TROHDW and TROIH denotes type of roofs owned by the households which drink rainwater and owned by all the interrogated households, respectively, while GCS denotes galvanised corrugated sheet)

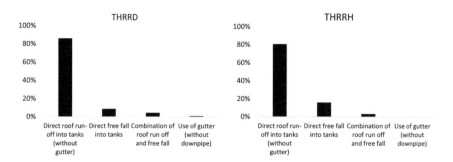

Figure 7.7 Techniques for harvesting rainwater by respondent (THRRD and THRRH denote Techniques of harvesting by rainwater drinkers and Techniques of harvesting by rainwater harvesters, respectively)

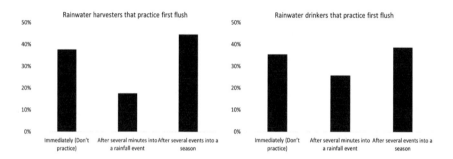

Figure 7.8 Percentage of rainwater harvesters/drinkers that practice first flush

This study findings revealed that none of the households had automated first flushing mechanisms, but some of them had adopted various first flush practices. Some collected the rainwater minutes after a rainfall event, while others started rainwater collection after several rainfall events during the rainy season. First flush is especially important for those who drink rainwater since it helps to reduce contamination from roof tops.

This study's survey findings also show that the neighbourhood's cleanliness and hygiene are generally poor. Of the 125 households that took part in the survey study, 90% (113) stated that they dispose of their greywater on the ground around their building because there are no reliable drainage systems in place. This promotes greywater evaporation and downward infiltration into the groundwater. Both mechanisms are likely to leave a residual solid waste within the first few millimetres of soil. Furthermore, for most households, the distance between their rainwater storage vessel and toilet is generally less than 10 m, thereby suggesting the probability of microorganism transmittance. Also, it was noted that boreholes used to deliver potable water were frequently constructed near septic tanks, which increases the danger of contamination, particularly in areas with a shallow water table.

Poorly designed septic tanks can pollute the water table in an area. Pathogens, faecal bacteria, phosphate, nitrogen, organic matter, suspended particles, pharmaceutical compounds, household detergents, and other contaminants that pose risks to freshwater resources are all present in septic tank effluent. Water samples can be examined for elements originating from septic systems to determine the quality of drinking water from shallow domestic wells that may be impacted by seepage from septic systems. Many tracers have been used in previous studies to show how septic systems affect water from domestic wells, including bacteria, viral indicators, dissolved organic carbon, nitrogen species, nitrogen, and boron isotopes, as well as organic substances including prescription and over-the-counter medications. Shallow domestic wells and narrow aquifer wells with sand points fewer than approximately 14 m in depth appeared to be most susceptible to septic waste contamination (Verstraeten et al., 2004; The James Hutton Institute, 2019).

7.3.2 Development of QMRA Model

The average pathogen densities, the average water consumption for exposure scenarios, the dose–response relationships for pathogens, and the conditional probability of illness are all necessary information to generate a QMRA model to characterise the *E. coli* exposure risk (Gerba et al., 1996; Hunter et al., 2003; Whelan et al., 2014). The results from the various HHTTs scenarios in Ikorodu are shown in Table 7.1 together with the estimated pathogenic *E. coli* doses that were consumed by the residents. Prior to applying various HHTTs, the amount of pathogenic *E. coli* detected using the Colilert method across 49 rainfall events was determined to be 1.55 MPN/100 mL; however, after applying alum, the number reduced to 0.78 MPN/100 mL (Table 7.1). With other applied HHTTs, such as boiling, chlorine, and alum and chlorine, no *E. coli* trace was found (as boiling and chlorine operate as effective disinfectants).

Equations 7.1 and 7.2 were used to determine the probability $P(i)$ and numbers N for various scenarios, and the results are shown in Table 7.2.

Table 7.1 Estimated dosages of pathogenic *E. coli* from various HHTTs

S/N	HHTTs applied	E. coli Dose (MPN/100mL)
1	Boil	0
2	Alum	0.78
3	Chlorine	0
4	No HHTT	1.55
5	Alum + Chlorine	0

Table 7.2 Probabilities and numbers of infection per 10,000 rural Ikorodu persons for different scenarios per single event

S/N	HHTTs applied	E. coli dose in 2000 mL (MPN)	P(i) %	N
1	Boil	0	0	0
2	Alum	15.54	0.0118	0.118
3	Chlorine	0	0	0
4	No HHTT	31.08	0.0234	0.234
5	Alum + Chlorine	0	0	0

The dose–response phase is depicted in the Figure 7.9, which evaluates the likelihood of a response by given a known dose (number of microorganisms) of the target pathogen. It demonstrates the relationship between the administered dose and the likelihood of infection in the exposed population. The findings from Tables 7.1 and 7.2 demonstrate the importance of using a disinfectant to eliminate harmful *E. coli* in the harvested rainwater. The calculated probability of infection and the doses of pathogenic *E. coli* have been found to have an exponential relationship (i.e., an increase in the consumed *E. coli* dosages leads to a significant increase in the number of infection).

The amount of Ikorodu residents exposed to pathogenic *E. coli* per 10,000 people was estimated in the final and fourth phase of risk characterisation. The survey results in Figure 7.10 were used to estimate the percentage of people who are exposed to hazardous *E. coli*. The survey estimated that 91% of all the households (i.e., 124, 678 out of the total 137, 009 households) harvested rainwater. Additionally, the figure illustrates the adjusted proportion of households that used various HHTTs. According to the calculations in

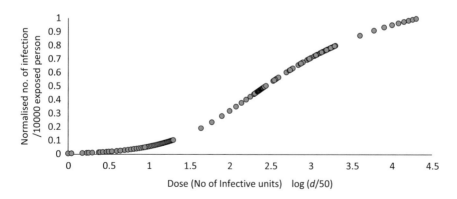

Figure 7.9 Beta-Poisson dose-response relationship. The dose-response relationships correlate N to P(i)

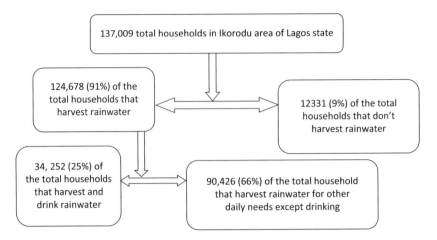

Figure 7.10 Estimation of the proportion of households using rainwater (the percentage in the chart were estimated from the survey outcome)

Table 7.3, 34,252 households in the Ikorodu area collected and drank rainwater during the rainy season, but only about 5% of all households utilise any kind of HHTT. This suggests that a sizeable amount of households are at risk of exposure to the target pathogen found in rainwater storage tanks.

In the worst situation, *E. coli* can remain in the rainwater storage tank for the 281 days of the rainy season. This was applied to Equation 7.3 to produce the results in Table 7.4. Given that *E. coli* may not necessarily be present in the tank for the entire period, it is considered a worst-case scenario. Table 7.4 displays the maximum infection risk for people exposed to contaminated tank water. The findings revealed that within 10,000 households, the maximum risk is 96 for those who use alum and 100 for those who do not. This highlights the importance of applying an appropriate disinfectant.

Table 7.3 Survey results of different HHTTs (including no HHTT) on rainwater drinkers

S/N	HHTTs	% of household who drink rainwater to the total household population	Household population of rainwater drinkers
1	Boil	13.7	18,770
2	Alum	5.6	7,741
3	Chlorine	0.8	1,096
4	No HHTT	4.9	6,645
		25	34,252

Table 7.4 Risk of infection for people exposed to contaminated tank water for four scenarios

HHTTs	Maximum Ps (%)	Maximum Rs	Number of events per year	Maximum Py (%)	Maximum Ry
Boil	0	0	0	0	0
Alum	0.0118	118	281	0.00964	96
Chlorine	0	0	0	0	0
No HHTT	0.0234	234	281	0.00999	100
Alum + Chlorine	0	0	0	0	0

Notes: Ps/Rs and Py/Ry denotes the probability/range of infection risk per 10, 000 exposed households with rainwater tanks from single event and probability/range of infection risk per 10, 000 exposed households per year in the Ikorodu district, respectively.

7.3.3 Public Health Impact of Different Scenarios in the Ikorodu Area of Lagos

In Ikorodu, approximately 25% of all households drink rainwater, and 10.5% of all households are at risk of exposure to *E. coli*, according to the result analysis in Table 7.3. This estimate considers the 6645 households that do not apply any sort of treatment before drinking the collected rainwater, as well as the 7741 households that just use alum (Table 7.3). In more specific terms, 10.5% was obtained by dividing households that drink rainwater without disinfectants (i.e., total households that does not apply HHTT and apply only alum) by the total households in Ikorodu $\left(= \dfrac{(7741+6645)}{137009} \times 100\% \right)$. In short, among 34, 252 households that harvest and drink rainwater in the Ikorodu area of Lagos, 42% (or 14, 386) of them are exposed to the pathogen risk. This 42% estimate was obtained by dividing households that drink rainwater without disinfectants by households that harvest and drink rainwater in Ikorodu $\left(= \dfrac{(7741+6645)}{34252} \times 100\% \right)$.

The residents who apply only alum and those who do not use any kind of HHTT are recommended to use one form of disinfectants, i.e., chlorine, boiling, or a mix of chlorine and alum. As evident from this study's results, the chlorine and boiling treatment techniques almost completely eliminate the bacteria in the harvested rainwater, which alum alone is unable to do. The disinfectant nature of chlorine and boiling are responsible for the low or non-existent risk potential displayed by their respective scenarios (Table 7.4).

Although it has been shown that 91% of the respondents collect and use RHRW for everyday requirements, untreated RHRW can still be utilised for non-potable purposes (such as farming, cleaning, and bathroom uses). While using the QMRA model, Ahmed et al. (2010) calculated a negligibly low risk level of *E. coli* infection when the RHRW was used as non-potable water in Australia. Hamilton et al. (2017) estimated the risks to human health from potable and non-potable uses of roof-harvested rainwater for *Legionella spp.* and Mycobacterium avium complex using the QMRA model. Their findings indicated that there were significant risks associated with RHRW consumption, while other domestic and agricultural uses (such as washing cars and clothes, flushing toilets, and using RHRW for lettuce irrigation) were suggested as safe RHRW applications for all populations, including those who have compromised immune systems. The conclusion in this study corresponds with the results from Ahmed et al. (2010) and Hamilton et al. (2017).

7.4 DISCUSSION

Sustainable and sufficient drinking water sources for daily use, especially in economically less developed and under-developed countries, have been challenging to maintain. Rainfall events have the propensity to alleviate this challenge, especially in the rainy season. Since the literature shows that harvesting and drinking rainwater is a norm in less developed and under-developed regions in the world, it is imperative to investigate the health risk involved in using or consuming rainwater (Lester, 1992; Brodribb et al., 1995; Merritt et al., 1999; Simmons et al., 2001; Ashbolt and Kirk, 2006; Ahmed et al., 2010; John et al., 2021d). The survey's results revealed a significant reliance on rainwater, as 114 of the 125 respondent households collected RHRW throughout the rainy season. Further, 31 out of those 114 households of rainwater harvesters drank rainwater. Detailed analysis also disclosed that 17 out of those 31 households boiled the collected rainwater before consuming it, while only one used chlorine treatment. In contrast, the remaining households (7) used solely alum, which is a coagulant rather than a disinfectant. The QMRA results evidenced that the load level from the untreated or alum-treated scenarios in Table 7.2 greatly surpassed the reference threshold set by the World Health Organization, where this level can be significantly lowered by disinfecting rainwater (i.e., by boiling, using chlorine, or a mixed approach using both chlorine and alum).

To illustrate a worst-case scenario, this study employed a 7% estimate for viable and dangerous *E. coli* (after considering suggestions by Ahmed et al., 2010; Machader et al., 2013; and John et al., 2012c). A risk of infection from ingesting untreated rainwater contaminated with *Giardia lambia* and *Salmonella spp.* was also demonstrated by Ahmed et al. (2010). For *Giardia lambia* and *Salmonella spp.*, the maximum risk per 10,000 people per year is 65 and 54, respectively. Another QMRA study evaluated the health risk of drinking RHRW that contains *Campylobacter*, and the results showed a risk of 3.4 per 10,000 people per year, while the risk of infection through domestic uses, washing, and bathing is very low (Hora et al., 2018).

The fact that *E. coli* is a surrogate parameter closely related to bacterial contamination, whether in the form of a heterotrophic or faecal group, has to be studied and proven. In some situations, particularly for people who are immunosuppressed, *E. coli* itself has the potential to be dangerous. This study estimated the maximum 7% of potentially dangerous *E. coli* strains, where it represents a necessary conservative approach, especially when considering bottom-line circumstances. Furthermore, due to its proven potential to produce a larger risk number, *E. coli* can be viewed as one of the best choices for evaluating the worst risk possibility. This was supported by a comparison of findings from recent and earlier research, which revealed varying degrees of risk burden from diverse diseases (Ahmed et al., 2010).

Over the years, no epidemic or instance of *E. coli* infection from any source has been documented in this study region, even though this may be attributed to its lack of records. Across the globe, a number of illnesses and pandemics linked to drinking rainwater have been reported (Brodribb et al., 1995; Merritt et al., 1999; Simmons et al., 2001; Ashbolt and Kirk, 2006; Ahmed et al., 2010). The QMRA is one of the most recognised and frequently used approaches to estimate the health risk of consuming water from any source, despite some uncertainties or assumptions made during the assessment of health risks using the QMRA. Due to the assumption that all *E. coli* cells are alive and contagious, which is not always true, the risk evaluated by the QMRA technique may have been overestimated. Apart from this overestimation, the QMRA results strongly indicate that sterilising collected rainwater to eliminate or minimise the risk of infection is crucial, especially before drinking.

This study's QMRA suggested using more roof-harvested rainwater in the rural Ikorodu district of Lagos together with at least one type of disinfection method. Among numerous ways to sterilise drinking water (such as the use of chlorine, sodium hypochlorite, chlorine dioxide, hydrogen peroxide, bromine, copper/silver ionisation, ultraviolet radiation, chloroamines, and boiling), this study has examined and chosen boiling, chlorination, and a mix of alum and chlorine due to their effectiveness and their popularity and relevance in the study region. Additionally, it is critical that more research on the long-term public health risk associated with rainwater consumption is conducted in this study area. This is because the results from the conducted survey demonstrated that 25% of the residents in Ikorodu use roof-harvested rainwater as their primary source of drinking water during the rainy season, and an additional 66% of the residents use the harvested rainwater for other daily activities (Figure 7.3).

Another topic for discussion is the origins of contamination in rainwater storage tanks. A properly managed and maintained supply system contributes

to further preservation of water quality. If the sources of pollution arc climi
nated and the rainwater system is kept in good working order, rainwater
can be a reliable and safe source of drinking water. For example, contaminants from roofs and gutters can be washed into rainwater storage tanks
after rainfall events, causing the collection to become contaminated. Varied
locations have different major sources of rainwater contamination, which
can range from rodents to dust, leaves, smoke, pollen, pesticide sprays, ferti-
lisers, waste, birds, small animals, and insects. Hazardous substances including asbestos, lead, and copper can also get into the water through roofing
materials, gutters, pipelines, and storage materials.

To prevent breeding of mosquitoes and further contaminants getting into
rainwater storage tanks, they should always be covered. The cleanliness and
hygiene of the roofs and rainwater storage tanks play critical roles in the collected water quality. It is essential to do routine checks on the various components used to collect rainwater. Simple modifications, such as a roof gutter
that is enclosed, a mesh trap, a first flush diversion device in the downpipe,
or a sump with fine mesh, can aid in keeping leaves and other debris out.
Constant maintenance, including keeping large tree branches off the roof;
ensuring the inlet and overflow screens are secure, in good condition, and
clean; cleaning first flush devices after rainfall events; and frequently inspect-
ing rainwater tanks for the build-up of sludge, is also crucial (Ahmed et al.,
2011; CDC, 2021; John et al., 2021d; Department of Health, 2022).

7.5 CONCLUSION

Using the QMRA analysis, this chapter examined the health risk associated with the use of RHRW in the Ikorodu neighbourhood of Lagos State,
Nigeria. The survey data collected from 125 carefully chosen representative
households revealed that rainwater is one of the alternative sources of drinking water during the rainy season, even though it could be contaminated with
pollutants, microbes, heavy metals, or harmful substances. The percentage of
people that consumed rainwater was about 25. The majority of those who
use rainwater for potable purposes have also used one of the HHTTs that
have been examined (such as chlorine, boiling, or a combination of chlorine
and alum treatments), but a minority group of households continue to drink
untreated rainwater. Using the guideline for drinking water quality (WHO,
2011; George et al., 2015), the risk associated with RHRW ingestion was
further assessed. The results showed that the pathogenic *E. coli* exposure
risk from consuming RHRW without application of any HHTT and with
application of only alum was 100 and 96, respectively, per 10,000 exposed
households per year, which shows the importance of applying disinfectant
to the harvested rainwater. Finally, any other similar under-developed community or nation that experiences rainwater contamination may also find the
presented results and guides useful.

Chapter 8

Concluding Remarks and Recommendations for Further Work

ABSTRACT

A lack of clean useable water, induced by water scarcity and contamination, is a serious global dilemma that affects billions of people. Water scarcity is a result of both natural and anthropogenic factors. Fresh water is considered abundant enough to provide for all seven billion people on the earth, but it is dispersed unevenly, polluted, and badly managed. Rainwater is a sustainable water source that can bridge scarcity, especially in tropical areas where intensity is high. The presence of physical and microbial contamination has been the main challenge in relation to drinking rainwater. Field measurements together with the analyses performed and discussed in this book aim to investigate different mechanisms used to improve rainwater quality, which include household water treatment, sedimentation, and first flush techniques. The percentage of microbes associated with the solids was also assessed, where their risk to human health has been analysed with Quantitative Microbial Risk Analysis (QMRA). This chapter concludes this book's key findings and recommends further works to compliment current research.

8.1 INTRODUCTION

Globally, 2.2 billion people do not have consistent access to regulated and safe drinking water (WHO and UNICEF, 2019). In terms of the hygienic sector, 2.3 billion people, or 1 in 3 people, lack access to simple hand-washing facilities at home (UNICEF and WHO, 2021; Wellaware, 2023). Sub-Saharan Africa has the largest number of water-stressed countries in the world, as it is home to around 40% of individuals without an adequate source of drinking water (UN, 2014; UNICEF and WHO, 2021). Unsafe water, sanitation, and hygiene (WASH) can have fatal effects for children. Due to a lack of suitable WASH services, more than 297,000 children under 5 die each year (approximately 814 children per day) from diarrheal infections as recorded in 2019 (WHO/UNICEF, 2019), which was improved to 255, 500 children

DOI: 10.1201/9781003392576-8

(approximately 700 children per day) by 2021 (UNICEF and WHO, 2021). Diseases including cholera, typhoid, dysentery, diarrhoea, and polio can be spread by contaminated water.

Rainwater can contribute to mitigate water scarcity. In locations where pipe-borne and purified community water is not available, rainwater collection by individuals is one of the least complex, sustainable green technologies that yields a significant return from a low-cost investment. Also, rainwater collection systems encourage building owners to take responsibility for their own water supply, and the process educates them about water scarcity and the characteristics of various water sources (Lim and Jiang, 2013). Due to the limited studies, assistance, and financial support supplied by the appropriate authorities, the potential of rainwater has not yet been maximised (George et al., 2015). In rural places where it is expensive and challenging to get alternative sources of water, using rainwater as a potable source, particularly during the rainy season, is a common practice (John et al., 2021d). The findings of this study and previous literature disclose that although many people in under-developed and developing countries rely on rainfall as an alternative supply of drinkable water, it generally has poor microbiological quality (Crabtree et al., 1996; Simmons et al., 2001; Lye, 2002; Ahmed et al., 2010; John et al., 2021d).

8.2 SUMMARY, CONCLUSION, AND RECOMMENDATIONS

Chapter 1 introduced the background and the structure of the book while Chapter 2 reviewed the existing literature on rainwater quality determination techniques with a focus on developing and under-developed countries. Chapter 3 explored the different experimental methods used obtain the results in this study. The first experimental section of this book investigated the rate of atmospheric deposition on a single building at Ikorodu (the selected study site), and its variation over both rainy and dry seasons. Furthermore, local environmental factors (which include wind speed, rain intensity, and dry antecedent days) were analysed during both seasons to determine which of these are influential to depositions. The range of the bacteria attached to particles was also determined (i.e., Chapter 4). The second experimental section investigated the quality of roof-harvested rainwater (RHRW) from different depths of storage tank and the impact of sedimentation on improving the physical and microbial quality of roof-harvested and stored rainwaters. The impact of different household water treatment techniques on the rainwater was also studied (i.e., Chapter 5). Chapter 6 further analysed the impact of first flush practices on the quality of roof-harvested rainwater in the study area. Finally, Chapter 7 estimated the possibility of risk infection for residents who drink rainwater in the Ikorodu district of Lagos state and analysed the results from an administered questionnaire survey conducted in the study region.

The summary of findings from this book is highlighted below:

- In Chapter 4, it was found that the deposition of particulate matter from a single building showed that the total and net depositions (g/m²/week) in the dry season ranged from 15.21 to 27.55 and 1.171 to 2.167 respectively, with week 5 and 14 presenting the highest and lowest deposition. Furthermore, the total and net depositions of the particulate matter (g/m²/day) in the rainy season ranged from 0.38 to 1.90 and 0 to 0.77 respectively. As a result, it was observed that total depositions were higher during the dry season than during the rainy season. The higher total depositions during the dry season were a result of the much larger concentration of dry particles in the atmosphere during that time. This means that there was an accumulation of particles with a longer dry antecedent period and reduced cloudiness during the dry season. The outcomes from the deposition of particulate matter (in weeks 5 and 6 of Figure 4.5) further illustrated that the Harmattan period in the dry season gave the highest deposition. The fractionation (or serial filtration) of the total and net deposition showed that most of the sample sizes ranged between 10 μm and 100 μm. The results from the fractionation of the samples demonstrated that the percentage of solids (from the total and net deposition experiments) deposited in the 500 μm mesh filter during the dry season ranged from 5.2 to 9.2 and 4.7 to 8.1, respectively, while that retained by the 10 μm filter ranged from 69 to 75.1 and 69.8 to 77.2, respectively. In the rainy season, the fractionation of samples showed that the percentage of the solids (from the total and roof-runoff net deposition experiments) deposited in the 500 μm mesh filter ranged from 2.7 to 4.6 and 1.7 to 8.7 respectively, while that sieved by the 10 μm filter ranged from 67.6 to 71.3 and 44.9 to 75.8 respectively. Further, the results showed that not all the solids in the raw samples were accounted for after the serial filtration. This could be attributed to some of the solids being smaller than the 10 μm filter. The microbial analysis in the fractionation tests found that few or no bacteria (i.e., 0%–2%) was removed after passing the sample through the 500 μm filter mesh, and between 4% and 7% was eliminated after sieving with a 100 μm filter. There was increased removal (i.e., 10%–45%) after using 10 μm filter paper, whereas some bacteria were still found after the sample had been passed through all the filters. Fractionation results evidenced that the 10 μm filter paper removed the majority of the bacteria during the serial filtration process, while the 500 μm filter mesh removed few or no bacteria. The deposition experiments reported in this book were conducted only in residential areas because rainwater consumption mostly occurs there.
- Simple and multiple linear regression was used for statistical analysis in this book. Results from the multiple linear regression indicated that the

parameters influencing the deposition of particulate matter in both rainy and dry seasons include the amount of rain, wind speed, and period of dry antecedent days. The most significant factors in the dry and rainy seasons were wind speed and rainfall intensity respectively. The results of the simple linear regression proved that there is a significant positive correlation of 0.925 between the net depositions and exposure time. This illustrates that deposits accumulate over time (particularly during the dry season), with the wind speed and length of the dry antecedent days having the biggest impact on the deposits. Analysis of the results further quantified the common belief that there was more accumulation of deposit in the dry season than the rainy season.

- The results from Chapter 5 showed that the quality of roof-harvested rainwater is poorer than harvested free-fall rainwater, and that the quality of the rainwater harvested in the dry season is poorer than that harvested in the rainy season. This is due to the shorter period of dry antecedent days in the rainy period. However, neither type of harvested rainwater met the World Health Organisation's (WHO) guidelines for drinking water. As a result, it is recommended that one form of household water treatment technique is applied. The impact of the sedimentation process in RHRW storage tanks was investigated for both physical and microbial parameters of water. Changes in the investigated microbial parameters (i.e., total coliform and *E. coli*) and the physical parameters (especially turbidity and TSS) in the tank for the different rainfall events disclosed that the quality of water from the top two levels improved with time, while that of the bottom deteriorated with time. The changes in the parameters were attributed to the settling of bacteria attached to solids. The fractionation experiments further illustrated that the significant fractions of the bacterial indicator organisms are associated with settle-able particles in the rainwater storage tank. The outcomes from the experimental procedures showed that sedimentation of particles in the storage tank provided a reasonable means for separating free phase organisms and those attached to less dense particles from settle-able particles. The enumerated bacteria were also showed to increase downwardly at the different depths in the tank, with the top and bottom level of the tank having the least and highest enumerated bacteria respectively. Hence, it is concluded that a significant portion of the bacteria are attached to solids. In Chapter 5, the household rainwater treatment techniques employed by the respondents of the administered questionnaires were further assessed. The three different household water treatment techniques applied to roof-harvested rainwater were alum, boiling, and chlorination. Results showed that boiling and chlorination were more effective in eliminating bacteria, while alum was more effective for reducing turbidity and solids in the water. Thus, it is recommended that a combination of chlorine and alum should be applied to stored rainwater, where their combined method has also been investigated.

- Chapter 6 evaluated the impact of the first flush practice on the quality of collected rainwater. Every first flush trial showed at least a 40% reduction in pollution load. This study also found that corrugated galvanised roofing provided the most superior microbiological quality for RHRW when compared to other tested materials (which included plastic, asbestos, and aluminium roofing materials) . First flush has a favourable impact on captured rainwater, but it does not completely remove impurities. No matter how much rainwater was diverted, it was discovered that the quality was poor because the physical and microbial parameters examined did not meet the WHO's guidelines for drinking water. It is therefore advised that household water treatment techniques must be used to disinfect the rainwater.

- In Chapter 7, the analysis of the administered questionnaires showed that boreholes are the main source of drinking water throughout the year. The importance of rainwater is highlighted by the fact that the proportion of respondents who utilise boreholes reduced from 82% in the dry season to 55% in the rainy season (see Figure 7.3). For the projected number of infection risks per 10,000 exposed households annually, the QMRA model estimated the highest *E. coli* exposure risk from consuming RHRW to be 96 and 100 for those who apply only alum and no HHTT, respectively. The QMRA was based on the assumption that 7% of *E. coli* are viable and dangerous. Since results analysis of the rainwater showed that the quality of rainwater is poor, it is imperative that one form of household water treatment technique is applied, especially for drinking water. Also, it is advised that those who collect rainwater install first flush devices or practice first flush, maintain proper gutter and roof hygiene, and eliminate overhanging tree branches and other objects to prevent the deposition of additional particulate matter and animal faeces to facilitate higher quality roof-harvested rainwater. The disinfection techniques are imperative considering that some groups of the population have impaired immune systems, such as children under the age of 5 and seniors over 60. However, the only target pathogen used for the development of the QMRA in this book was *E. coli*. It is therefore recommended that more pathogens could be used to develop the QMRA in under-developed and developing countries. This study also determined the percentage of the settle-able *E. coli* that are attached to solids; more studies on serial filtration of other microorganisms that are attached to the solids need to be done.

- The roof area of the surveyed 125 households ranging from 42 m² to 164 m², with an average of 92 m² per household. This implies that the average roof area is approximately 18 m² per person. This data, alongside with information on rainfall patterns and population density, can be used to assess the ability of a rainwater system to meet the local water demand. Governmental and non-governmental organisations and other interest groups are encouraged to use this information to invest in public rainwater collecting systems to help meet water demand, particularly during the rainy season.

This study estimated the health risk from the ingestion of RHRW in the Ikorodu area of Lagos state, with resultant findings strengthening the need for professional attention to the microbial contamination of water. It is imperative to develop and implement a sustainable water safety plan for harvesting rainwater in Nigeria so that long-term epidemiology studies can be easily carried out. Finally, adopting efficient, effective, and sustainable control techniques is advised for those who collect rainwater (such as providing safe drinking rainwater through adequate treatment combining chlorination and alum, appropriate health education focused on hygiene practices, sanitation, and periodic medical check-ups). These measures will educate rainwater harvesters on better practices for harvesting and treating rainwater, improving the quality of the consumed rainwater and thus improving the health of residents.

Appendices

Appendix A: General and Specific Definitions

GENERAL DEFINITION

Developing Countries or Less Economically Developed Countries

There is no particular definition of a *developing country*. This term is used to define areas in a wide context, because quoted research and statistics frequently discuss developing countries. An alternative definition is to liken the term *developing country* to a *less industrialised economy* (LIE), defined as a country or region with a "low capacity to design and manufacture original products" (Donaldson, 2006). In literature, they may be defined according to the World Bank as low- and middle-income countries or by the United Nations Development Programme's Human Development Index as countries with medium and low human development (UNDP, 2010; World Bank, 2013).

With respect to regions, Asia excluding Japan, Africa, the Americas excluding the Caribbean, Central America, Northern America, South America, and Oceania excluding Australia and New Zealand are considered developing countries by the United Nations (United Nations Statistics Division, 2010).

Total Coliform

These include bacteria that are detected in the soil and in water that has been affected by human or animal waste.

Escherichia coli (E. coli)

Escherichia coli is a facultatively anaerobic, gram-negative, rod-shaped bacterium of the genus *Escherichia* that is usually found in the stomach of warm-blooded organisms. This is the main species in the faecal coliform group. Of the five common groups of bacteria that make up the total coliforms, only *E. coli* is commonly not found growing and reproducing in the environment.

Therefore, *E. coli* is known to be the species of coliform bacteria that is the best indicator of faecal pollution and the possible presence of pathogens.

SPECIFIC DEFINITION

Fractionation of Samples by Serial Filtration

This method prepares sub-samples of suspended particulate matter within a given size range for subsequent gravimetric and/or microbiological analysis.

Net Deposition

The net deposition experiment determines the mass of particulate matter that is deposited on a rooftop by direct measurement. In this respect, net deposition refers to the equilibrium of deposition and re-suspension processes.

Total Atmospheric Deposition

The total atmospheric deposition method determines the mass of particulate matter that deposits from the air through a horizontal plane in each time period.

Roof Deposit Runoff

The roof runoff experimental method determines the mass of particulate matter removed from a roof in each rainfall event.

Significant: Statistical significance was determined using several statistical tests (ANOVA test, multiple linear regression analysis); more detail is given in the appropriate sections. Unless otherwise stated, $P < 0.05$ is used to denote significance.

Appendix B: Design of the Two Gutters Used for the Experiments (for Roof Runoff Deposition and Sedimentation Experiments)

Design of the First Gutter in the Sedimentation Experiment

Diameter of the tank = 60 cm = 0.6 m
  Height of the tank = 90 cm = 0.9 m

$$\text{The volume of the tank} = \frac{\pi d2H}{4} = \frac{(3.142 * 0.6 * 0.6 * 0.9)}{4} = 0.255 \text{ m}^3$$

Length of the gutter = 2.5 m
Dimension of the gutter catchment used for the sedimentation experiment = 2.5 m × 4 m

After simulation of the rainfall in a year in accordance with Thomas (2002), the maximum intensity is approximately 50 mm/day = 2.08 mm/hour (0.00208 m/hour).

Since the expected maximum intensity of the rainfall is approximately 0.00208 m/hour, thus for a 3- and 4-hour rain event, the intensity becomes 0.00624 m/hour and 0.00832 m/hour, respectively.

$$\text{Thus, time to fill the 0.255 m}^3 \text{ tank} = \frac{\text{Volume}(m^3)}{\text{Quantity}(m^3/\text{hour})} =$$

$$\frac{0.255\,(m3)}{(2.5m\,X4m\,X0.00624\,(m/\text{hour}))} = 4 \text{ hours}$$

$$\text{Thus, time to fill the 0.255 m}^3 \text{ tank} = \frac{0.255\,(m3)}{(2.5m\,X4m\,X0.00832\,(m/\text{hour}))}$$

= 2.43 hours = 3 hours
  Thus a 2.5 m gutter will be used.

Design for the second gutter for roof runoff deposition

Diameter of the tank = 60 cm = 0.6 m
  Height of the tank = 90 cm = 0.9 m

  The volume of the tank = $\dfrac{\pi d2H}{4}$ = $\dfrac{(3.142 * 0.6 * 0.6 * 0.9)}{4}$ = $0.255m^3$

  Length of the gutter = 0.5 m
  Dimension of the gutter catchment used for the sedimentation experiment = 0.5 m × 4 m
  After simulation of the rainfall in a year in accordance with Thomas (2002), the maximum intensity is approximately 50 mm/day = 2.08 mm/hour (0.00208 m/hour).
  After simulation of the rainfall in a year in accordance with Thomas (2002), the maximum intensity is approximately 50 mm/day = 2.08 mm/hour (0.00208 m/hour). Since the expected maximum intensity of the rainfall is approximately 0.00208m/hour, thus for a 4-hour rain event, the intensity becomes 0.00624 m/hour.
  Thus, time to fill the 0.255 m^3 = $\dfrac{0.255m^3}{(4m \times 0.5m \times 0.00832m / hour)}$ = 15 hours.
  This length of gutter will take over a day, therefore making it easy to be completely mixed.
  Atmospheric deposition (g/m²/week) =
$$\dfrac{(\text{mass of filter} + \text{particles}) - \text{mass of filter } (g)}{Area(\text{width of roof X length of gutter})m^2} \text{ X } \dfrac{1}{Week}$$

  For Gutter 1 (i.e., Sedimentation analysis), the dimension is 2.5 m × 0.2 m × 0.1 m.
  For Gutter 2 (i.e., accumulation analysis), the dimension is 0.5 m × 0.2 m × 0.1 m.

Construction of Gutter

Aluminium box gutters were fabricated for this research. These were chosen because of their ability to withstand higher intensities of rainfall. The dimensions of the gutter were designed for different purposes. The gutter was properly fitted to the fascia board via screws/nails, and the storage vessel was kept at the end of the gutter down-pipe to collect the roof-harvested rainwater. The aluminium box gutter will be designed as shown. The diameter of the pipe is 50 mm. The gutter was attached to the fascia board such that the higher side of the width was slightly above the roof so that accumulations of rooftop depositions would not spill over.

Appendix C: Experimental Results from the Property in the Rainy Season

S/N	Date of rain event	DAD	Amount of rainfall by rain gauge (inch/day)	Volume of rain harvested (m³)	Roof area (4m × 0.5m)	Mass of solids deposited (g)	Left-over deposits on tiles after rain events (g)	Roof runoff net deposit (RRND) (g/m²)	Total roof runoff net deposit (TRRND) (g/m²)	Total deposition (g/m²)	Wind (MPH)
1	03/04/2015	3	0.15	0.011	2	0.138	—	0.069	0.13	5.71	1.56
2	05/04/2015	2	0.06	0.003	2	0.054	0.001	0.029	0.02	2.82	4.05
3	07/04/2015	2	0.71	0.073	2	0.059	—	0.029	0.53	2.76	2.17
4	09/04/2015	2	0.43	0.047	2	0.048	—	0.024	0.28	2.58	2.04
5	10/04/2015	1	0.08	0.011	2	0.039	—	0.02	0.11	1.16	2.38
6	11/04/2015	1	0.03	0.003	2	0.034	0.003	0.023	0.03	1.01	1.99
7	12/04/2015	1	0.39	0.032	2	0.036	—	0.018	0.29	1.04	0.7
8	24/04/2015	12	0.48	0.05	2	0.438	—	0.219	0.46	19.85	2.57
9	25/04/2015	1	0.02	0.018	2	0.018	0.014	0.038	0.34	0.99	2.81
10	28/04/2015	3	0.11	0.0049	2	0.118	—	0.059	0.05	5.23	1.97
11	29/04/2015	1	0.03	0.004	2	0.025	0.007	0.027	0.05	0.95	2.19
12	01/05/2015	2	1.12	0.073	2	0.072	—	0.036	0.66	2.7	4.5
13	03/05/2015	2	0.15	0.019	2	0.063	—	0.032	0.15	2.61	3.51
14	04/05/2015	1	0.09	0.014	2	0.035	—	0.018	0.13	1.22	2.5
15	12/05/2015	8	0.42	0.048	2	0.302	—	0.151	0.45	13.3	2.26
16	13/05/2015	1	0.3	0.028	2	0.031	—	0.016	0.22	1.4	2.56
17	17/05/2015	4	0.32	0.037	2	0.145	—	0.073	0.34	5.8	0.82
18	19/05/2015	2	0.05	0.002	2	0.041	0.017	0.056	0.03	2.49	1.59
19	28/05/2015	9	0.13	0.0026	2	0.384	—	0.192	0.03	14.29	4.19
20	29/05/2015	1	0.43	0.053	2	0.029	—	0.015	0.4	1.26	1.99
21	31/05/2015	2	0.07	0.0021	2	0.054	—	0.027	0.01	2.25	3.77
22	02/06/2015	2	1.44	0.085	2	0.069	—	0.035	0.74	2.39	3.34
23	03/06/2015	1	0.72	0.059	2	0.018	—	0.009	0.27	0.93	1.21
24	04/06/2015	1	0.37	0.048	2	0	—	0	0	0.61	1.44
25	05/06/2015	1	0.18	0.024	2	0.024	—	0.012	0.14	0.67	2.57

S/N	Date of rain event	DAD	Amount of rainfall by rain gauge (inch/day)	Volume of rain harvested (m³)	Roof area (4m × 0.5m)	Mass of solids deposited (g)	Left-over deposits on tiles after rain events (g)	Roof runoff net deposit (RRND) (g/m²)	Total roof runoff net deposit (TRRND) (g/m²)	Total deposition (g/m²)	Wind (MPH)
26	07/06/2015	2	0.16	0.021	2	0.035	—	0.018	0.09	1.44	2.86
27	08/06/2015	1	0.23	0.026	2	0	—	0	0	0.51	2.72
28	09/06/2015	1	0.87	0.067	2	0.026	—	0.013	0.44	0.63	2.72
29	14/06/2015	5	1.03	0.0081	2	0.189	—	0.95	0.77	8.11	3.74
30	15/06/2015	1	0.1	0.0032	2	0.024	—	0.012	0.02	0.47	5.61
31	20/06/2015	5	1.03	0.045	2	0.194	—	0.97	4.37	4.97	5.89
32	22/06/2015	2	0.49	0.059	2	0.073	—	0.037	0.55	1.35	1.54
33	28/06/2015	6	0.76	0.064	2	0.202	—	0.101	0.54	5.33	6.28
34	29/06/2015	1	0.15	0.022	2	0.021	—	0.011	0.12	0.45	5.51
35	30/06/2015	1	0.26	0.028	2	0	—	0	0	0.49	6.91
36	02/07/2015	2	0.64	0.069	2	0.032	0.005	0.016	0.28	1.14	5.76
37	03/07/2015	1	0.07	0.002	2	0.011	0.008	0.016	0.02	0.41	2.19
38	04/07/2015	1	0.02	0.0021	2	0.009	0.012	0.021	0.02	0.38	1.23
39	05/07/2015	1	0.06	0.021	2	0.006	—	0.028	0.29	0.43	0.73
40	15/07/2015	10	0.12	0.0036	2	0.263	—	0.132	0.02	13.48	4.27
41	16/07/2015	1	1.03	0.052	2	0.014	—	0.007	0.18	0.39	2.24
42	20/07/2015	4	0.06	0.0022	2	0.043	0.018	0.059	0.02	4.16	8.96
43	25/07/2015	5	0.02	0.0075	2	0.025	0.029	0.073	0.05	4.48	5.83
44	27/07/2015	2	0.55	0.046	2	0.034	—	0.017	0.2	1.01	7.99
45	30/07/2015	3	0.15	0.016	N/D	N/D	N/D	N/D	N/D	N/D	6.06

Note: $TRRND = \dfrac{(RRND * \text{volume of the harvested rainwater} * 1000\ L)}{(\text{volume of analysed rainwater (i.e., } 2\ L) * DAD)}$

Appendix D: Fractionation Results from Both Seasons

Table D1 Fractionation of total and net deposition and its associated enumerated bacteria in the dry season

Filter size	S/N	Ma (g)	Mb (g)	TC1a (MPN /100 mL)	TC1b (MPN /100 mL)	TC2a (MPN /100 mL)	TC2b (MPN /100 mL)	TCa diluted (MPN /100 mL)	TCb diluted (MPN /100 mL)	EC1a (MPN /100 mL)	EC1b (MPN / 100 mL)	EC2a (MPN / 100 mL)	EC2b (MPN /100 mL)	ECa diluted (MPN /100 mL)	ECb diluted (MPN / 100 mL)
Raw sample	1	0.779	0.552	101.3	65.9	101.3	65.9	32.4	12.4	19.2	11.1	20.7	11.1	8.7	5.3
Raw sample	2	0.836	0.652	118.4	73.8	118.4	73.8	36.4	16.4	23.8	13.7	23.8	13.7	9.9	6.4
Raw sample	3	0.947	0.729	144.5	94.5	144.5	94.5	30.6	19.2	28.8	17.8	28.8	17.8	12.4	9.9
Raw sample	4	0.894	0.697	101.3	83.1	101.3	83.1	23.8	16.4	19.2	13.7	19.2	13.7	6.4	6.4
Raw sample	5	0.818	0.593	83.1	65.9	83.1	65.9	20.7	16.4	17.8	11.1	17.8	11.1	8.7	6.4
Raw sample	6	0.784	0.586	73.8	62.4	78.2	62.4	17.8	15	15	12.4	15	12.4	7.5	7.5
Raw sample	7	0.772	0.563	53.1	40.6	53.1	40.6	12.4	11.1	9.9	8.7	9.9	8.7	6.4	5.3
500 μm	1	0.068	0.026	101.3	65.9	101.3	65.9	30.6	11.1	19.2	9.9	19.2	9.9	7.5	5.3
500 μm	2	0.076	0.038	118.4	69.7	118.4	69.7	34.4	15	23.8	11.1	22.2	11.1	9.9	6.4
500 μm	3	0.087	0.059	129.8	88.5	129.8	88.5	28.8	17.8	27.1	15	27.1	15	11.1	8.7
500 μm	4	0.062	0.049	88.5	73.8	88.5	73.8	22.2	17.8	17.8	12.4	17.8	12.4	5.3	5.3
500 μm	5	0.051	0.037	78.2	59.1	78.2	59.1	17.8	15	15	9.9	15	9.9	7.5	5.3
500 μm	6	0.048	0.032	73.8	62.4	73.8	59.1	16.4	13.7	13.7	11.1	15	11.1	6.4	6.4
500 μm	7	0.04	0.028	50.4	38.4	50.4	38.4	11.1	9.9	8.7	7.5	8.7	7.5	5.3	4.2
100 μm	1	0.082	0.053	101.3	65.9	101.3	65.9	30.6	11.1	17.8	8.7	17.8	9.9	6.4	4.2
100 μm	2	0.095	0.064	94.5	65.9	94.5	65.9	32.4	13.7	22.2	9.9	22.2	9.9	9.9	5.3
100 μm	3	0.119	0.093	118.4	78.2	118.4	78.2	27.1	16.4	25.4	13.7	25.4	13.7	11.1	8.7
100 μm	4	0.098	0.079	83.1	69.7	83.1	69.7	19.2	16.4	16.4	11.1	16.4	11.1	4.2	4.2
100 μm	5	0.079	0.062	69.7	56	69.7	56	16.4	13.7	13.7	8.7	13.7	8.7	6.4	4.2
100 μm	6	0.067	0.058	65.9	59.1	65.9	59.1	15	12.4	12.4	9.9	12.4	9.9	5.3	5.3
100 μm	7	0.06	0.054	45.3	34.4	45.3	34.4	8.7	7.5	7.5	5.3	7.5	5.3	4.2	3.1
10 μm	1	0.561	0.423	88.5	59.1	88.5	59.1	28.8	9.9	16.4	7.5	17.8	7.5	5.3	3.1
10 μm	2	0.603	0.486	83.1	56	83.1	56	30.6	11.1	19.2	8.7	19.2	8.7	8.7	5.3
10 μm	3	0.653	0.509	101.3	73.8	101.3	73.8	28.8	16.4	23.8	11.1	23.8	11.1	9.9	7.5

Filter size	S/N	Ma (g)	Mb (g)	TC1a (MPN /100 mL)	TC1b (MPN /100 mL)	TC2a (MPN /100 mL)	TC2b (MPN /100 mL)	TCa diluted (MPN /100 mL)	TCb diluted (MPN /100 mL)	EC1a (MPN /100 mL)	EC1b (MPN /100 mL)	EC2a (MPN /100 mL)	EC2b (MPN /100 mL)	ECa diluted (MPN /100 mL)	ECb diluted (MPN /100 mL)
10 µm	4	0.618	0.494	73.8	62.4	73.8	62.4	17.8	15	13.7	9.9	13.7	9.9	3.1	3.1
10 µm	5	0.606	0.458	59.1	47.8	59.1	47.8	12.4	8.7	9.9	6.4	9.9	6.4	4.2	3.1
10 µm	6	0.589	0.437	62.4	53.1	62.4	53.1	13.7	9.9	11.1	7.5	11.1	7.5	4.2	4.2
10 µm	7	0.564	0.43	40.6	30.6	40.6	30.6	6.4	5.3	5.3	3.1	5.3	3.1	2	1

Where a and b denote total deposition and net deposition respectively, while M_a and M_b denote the mass of solids deposited for total and net deposition respectively. TC and EC denote total coliform and E. coli respectively.

Table D2 Fractionation of total deposition and its associated enumerated bacteria in the rainy season

Filter size	S/N	M (g)	TC1 (MPN /100 mL)	TC2 (MPN /100 mL)	TC diluted (MPN /100 mL)	EC1 (MPN /100 mL)	EC2 (MPN /100 mL)	EC diluted (MPN /100 mL)
Raw sample	1	0.742	47.8	47.8	15	9.9	9.9	5.3
Raw sample	2	0.733	28.8	28.8	13.7	11.1	11.1	6.4
Raw sample	3	0.716	25.4	25.4	11.1	9.9	9.9	4.2
Raw sample	4	0.693	17.8	17.8	9.9	7.5	7.5	3.1
Raw sample	5	0.669	13.7	13.7	7.5	6.4	6.4	2
500 μm	1	0.034	45.3	45.3	13.7	8.7	8.7	4.2
500 μm	2	0.029	27.1	27.1	12.4	9.9	9.9	5.3
500 μm	3	0.027	23.8	23.8	11.1	8.7	8.7	4.2
500 μm	4	0.019	15	15	8.7	7.5	7.5	3.1
500 μm	5	0.018	13.7	13.7	7.5	6.4	6.4	2
100 μm	1	0.057	40.6	40.6	11.1	6.4	6.4	2
100 μm	2	0.051	23.8	23.8	9.9	7.5	7.5	4.2
100 μm	3	0.046	22.2	22.2	8.7	6.4	6.4	3.1
100 μm	4	0.042	11.1	11.1	6.4	6.4	6.4	2
100 μm	5	0.037	9.9	9.9	5.3	4.2	4.2	1
10 μm	1	0.529	34.4	34.4	9.9	4.2	4.2	< 1
10 μm	2	0.518	17.8	17.8	6.4	5.3	5.3	1
10 μm	3	0.495	19.2	19.2	7.5	4.2	4.2	1
10 μm	4	0.488	8.7	8.7	3.1	4.2	4.2	< 1
10 μm	5	0.452	5.3	5.3	2	1	1	< 1

TC and EC denote total coliform and E. coli respectively while 1 and 2 denote duplicate tests.

Table D3 Fractionation of roof runoff net deposit and its associated enumerated bacteria in the rainy season

Filter size	S/N	M (g)	TC1 (MPN /100mL)	TC2 (MPN /100mL)	TC diluted (MPN /100mL)	EC1 (MPN /100mL)	EC2 (MPN /100mL)	EC diluted (MPN /100mL)
Raw sample	1	0.138	27.1	27.1	9.9	5.3	5.3	2
Raw sample	2	0.059	19.2	19.2	6.4	< 1	< 1	< 1
Raw sample	3	0.438	22.2	22.2	7.5	8.7	8.7	4.2
Raw sample	4	0.302	25.4	25.4	8.7	9.9	9.9	4.2
Raw sample	5	0.145	16.4	16.4	6.4	6.4	6.4	3.1
Raw sample	6	0.384	34.4	34.4	9.9	8.7	8.7	4.2
Raw sample	7	0.069	15	15	5.3	6.4	6.4	3.1
Raw sample	8	0.035	15	15	5.3	< 1	< 1	< 1
Raw sample	9	0.189	30.6	30.6	11.1	7.5	7.5	4.2
Raw sample	10	0.194	9.9	9.9	5.3	3.1	3.1	1
Raw sample	11	0.202	11.1	12.4	6.4	4.2	4.2	2
Raw sample	12	0.263	15	15	6.4	7.5	7.5	4.2
500 µm	1	0.005	23.8	23.8	8.7	4.2	4.2	1
500 µm	2	0.001	17.8	17.8	5.3	< 1	< 1	< 1
500 µm	3	0.028	20.7	20.7	6.4	7.4	7.4	4.2
500 µm	4	0.021	23.8	23.8	7.5	8.7	8.7	3.1
500 µm	5	0.009	15	15	5.3	6.4	6.4	3.1
500 µm	6	0.028	32.4	32.4	9.9	8.7	8.7	4.2
500 µm	7	0.006	15	15	5.3	6.4	6.4	3.1
500 µm	8	0.002	15	15	5.3	< 1	< 1	< 1
500 µm	9	0.008	28.8	28.8	9.9	7.5	6.4	4.2
500 µm	10	0.011	9.9	9.9	5.3	3.1	3.1	1
500 µm	11	0.016	9.9	9.9	6.4	3.1	3.1	1
500 µm	12	0.011	15	15	6.4	7.5	7.5	4.2
100 µm	1	0.018	20.7	20.7	6.4	3.1	3.1	< 1
100 µm	2	0.014	15	15	4.2	< 1	< 1	< 1
100 µm	3	0.057	17.8	17.8	4.2	5.3	5.3	3.1

Filter size	S/N	M (g)	TC1 (MPN /100mL)	TC2 (MPN /100mL)	TC diluted (MPN /100mL)	EC1 (MPN /100mL)	EC2 (MPN /100mL)	EC diluted (MPN /100mL)
100 µm	4	0.034	20.7	20.7	5.3	6.4	6.4	2
100 µm	5	0.022	13.7	13.7	4.2	4.2	4.2	2
100 µm	6	0.051	28.8	28.8	8.7	7.5	7.5	3.1
100 µm	7	0.019	12.4	12.4	4.2	5.3	5.3	2
100 µm	8	0.01	12.4	12.4	4.2	<1	<1	<1
100 µm	9	0.031	25.4	25.4	7.5	5.3	5.3	3.1
100 µm	10	0.034	8.7	8.7	4.2	2	2	<1
100 µm	11	0.04	8.7	8.7	5.3	2	2	<1
100 µm	12	0.034	11.1	11.1	4.2	5.3	5.3	2
10 µm	1	0.099	15	15	3.1	1	1	<1
10 µm	2	0.028	11.1	11.1	1	<1	<1	<1
10 µm	3	0.297	12.4	12.4	2	3.1	3.1	1
10 µm	4	0.229	16.4	16.4	3.1	3.1	3.1	<1
10 µm	5	0.103	9.9	9.9	2	1	1	<1
10 µm	6	0.264	23.8	23.8	3.1	3.1	3.1	1
10 µm	7	0.031	8.7	8.7	3.1	2	2	<1
10 µm	8	0.021	8.7	8.7	2	<1	<1	<1
10 µm	9	0.136	17.8	17.8	4.2	2	2	<1
10 µm	10	0.139	5.3	5.3	1	<1	<1	<1
10 µm	11	0.144	4.2	4.2	1	<1	<1	<1
10 µm	12	0.139	6.4	6.4	2	2	2	<1

TC and EC denote total coliform and *E. coli*, respectively, while 1 and 2 denote duplicate tests.

Appendix E: Administered Rainwater Questionnaire

Date of visit:
Location/Address:
Property reference:

About your household

Number of people living in the house:
What age and sex are the people living in the house?

Person	Age	Sex
e.g., Person 1	*e.g., 21*	*Male/Female*

Are animals kept at the property? If so what kind and how many?

Animal	No.
e.g., Goat	*e.g., 2*

Source of water

What is your **main source** of drinking water during:
- The dry season
- The rainy season
  Options
  a. Bottled water
  b. Wells/Boreholes

c. Untreated surface water
d. Community standpipes
e. Piped treated water from water supply company
f. Rainwater
g. Other (describe)
   Which of the other (supplementary) sources do you use?
   Do they apply water use strategies? i.e., do they use less water when it starts to run out?
   If you use rainwater, do you apply any form of household treatment technique (HTT)?
   If yes, which type?
a. Solar disinfection
b. Boiling
c. Chlorination
d. Alum addition
e. Other

If rainwater is collected, what is it used for?

a. Clothes washing
b. Bathing
c. Drinking
d. Cooking
e. Other (describe)

If used for drinking, can you estimate how much is consumed per day? i.e., do you decant a certain volume for use inside?

Details of Rainwater Harvesting System – IF USED

How long have you had the rainwater harvesting system?

Collection and Storage

Describe in as much detail as possible how rainwater is collected, stored, and used at the property.
   The following questions can be used as a prompt with the people at the property. Remember that they may not be very highly educated and may not understand what you mean by first flush, collection vessel, and storage vessel.
   Do you use a collection vessel and/or a storage vessel?
   Do you put the collection vessels out before or after it starts raining?
   If after, how long after/how many rainfall events occur before you start collecting the water?
   How is the rainwater captured?
1. Free-fall water directly into the collection vessel

2. Roof runoff directly in the collection vessel
3. Roof runoff via a guttering system in the collection vessel
4. A combination of the above
   Does the storage vessel have a first flush system? If so how does it work, and what are its dimensions?
   What are the volumes and material of the vessels used for rainwater harvesting?
- Vessel – height, width, breadth, or diameter.
- Vessel material – plastic, metal
   How do you get the water out of the storage vessel (i.e., by using a jug or via a tap near the bottom)?
   Describe the storage vessel. Does it have the following? (if so, gather dimensions):
- Overflow pipe
- Outlet taps
- Lid/cover
- State of repair/age
- Cleanliness
   What is the working volume of the tank (i.e., the volume between the overflow and outlet)?
   What is the depth from the outlet tap to the bottom of the storage vessel?
   How are the vessels cleaned/maintained?
   How frequently are the vessels cleaned/maintained?
   Where is the storage vessel kept (i.e., in shade or direct sunlight)?
- Photograph ALL vessels and any pipes and guttering used to collect the water.
- Save each photograph with the property reference in the file name.
- Note the location of the storage vessel on the mapping.
- Note the location of the sanitary facilities (toilet, clothes washing, etc.) on the mapping.
- Inspect the tank – how much debris is accumulated in the bottom?

Property Roof

What are the **overall dimensions** of the roof?
- Length (m)
- Width (m)
   Over what **actual area** is the water captured (i.e., is it only captured on one side by guttering?)
- Length (m)
- Width (m)
   What is the roof material?
- Galvanised corrugated steel
- Aluminium

- Clay tiles
- Thatch
- Other (describe)
  Describe the state of repair of the roof (i.e., is it corroded, heavily corroded, covered in plants, moss, etc.? Are tiles broken? Is the roof fouled with animal excrement?).
  Are there any over-hanging branches or other perches by birds?
- Take photographs of the roof and reference the file to the property.
- Take photographs of any guttering system and fascia boards/roof eaves visible.

Optional Questions

What is the highest household education? No school, primary school, secondary school, primary degree, and above.
What is the range of the household income?
N0.00–120,000.00 (£0.00–440) per annum
>N120, 000.00–600,000.00 (>£440.00–2400) per annum
>N600, 000.00–1, 200, 000 (>£2400.00–4800) per annum
What type of toilet do you use?
What is the distance between the toilet and the storage vessel?
What is the distance between the wastewater/garbage storage and the storage vessel?

Appendix F: The Most Probable Number (MPN) Table Used for Colilert Results (IDEXX, 2016)

51-Well Quanti-Tray MPN Table

Number of wells giving positive reaction per 100 mL sample	Most Probable Number (MPN)	95% confidence interval	
		Lower	Upper
0	< 1	0.0	3.7
1	1.0	0.3	5.6
2	2.0	0.6	7.3
3	3.1	1.1	9.0
4	4.2	1.7	10.7
5	5.3	2.3	12.3
6	6.4	3.0	13.9
7	7.5	3.7	15.5
8	8.7	4.5	17.1
9	9.9	5.3	18.8
10	11.1	6.1	20.5
11	12.4	7.9	22.1
12	13.7	7.9	23.9
13	15.0	8.8	25.7
14	16.4	9.8	27.5
15	17.8	10.8	29.4
16	19.2	11.9	31.3
17	20.7	13.0	33.3
18	22.2	14.1	35.2
19	23.8	15.3	37.3
20	25.4	16.5	39.4
21	27.1	17.7	41.6
22	28.8	19.0	43.9
23	30.6	20.4	46.3
24	32.4	21.8	48.7
25	34.4	23.3	51.2
26	36.4	24.7	53.9
27	38.4	26.4	56.6
28	40.6	28.0	59.5
29	42.9	29.7	62.5

Number of wells giving positive reaction per 100 mL sample	Most Probable Number (MPN)	95% confidence interval	
		Lower	Upper
30	45.3	31.5	65.5
31	47.8	33.4	69.0
32	50.4	35.4	72.5
33	53.1	37.5	76.2
34	56.0	39.7	80.1
35	59.1	42.0	84.4
36	62.4	44.6	88.8
37	65.9	47.2	93.7
38	69.7	50.0	99.0
39	73.8	53.1	104.8
40	78.2	56.4	111.2
41	83.1	59.9	118.3
42	88.5	63.9	126.2
43	94.5	68.2	135.4
44	101.3	73.1	146.0
45	109.1	78.6	158.7
46	118.4	85.0	174.5
47	129.8	92.7	195.0
48	144.5	102.3	224.1
49	165.2	115.2	272.2
50	200.5	135.8	387.6
51	>200.5	146.1	infinite

Appendix G: Main Rain Event Characteristics during the Sampling Trials

Harvest events	Dry antecedent period (days)	Rain intensity (mm/h)
1	3	6
2	2	2.4
3	2	28.4
4	2	17.2
5	1	3.2
6	1	1.2
7	1	15.6
8	12	19.2
9	1	0.8
10	3	4.4
11	1	1.2
12	2	44.8
13	2	6
`14	1	3.6
15	8	16.8
16	1	12
17	4	12.8
18	2	2
19	9	5.2
20	1	17.2
21	2	2.8
22	2	57.6
23	1	28.8
24	1	14.8
25	1	7.2
26	2	6.4
27	1	9.2
28	1	34.8
29	5	41.2
30	1	4
31	5	41.2
32	2	19.6
33	6	30.4

Harvest events	Dry antecedent period (days)	Rain intensity (mm/h)
34	1	6
35	1	10.4
36	2	25.6
37	1	2.8
38	1	0.8
39	1	2.4
40	10	4.8
41	1	41.2
42	4	2.4
43	5	0.8
44	2	22
45	3	6
46	5	32.2
47	2	17.1
48	8	5.3
49	3	6.8

References

Aas, W., Alleman, L.Y., Bieber, E., Gladtke, D., Houdret, J.L., Karlsson, V. and Monies, C. (2009). Comparison of methods for measuring atmospheric deposition of arsenic, cadmium, nickel and lead. *Journal of Environmental Monitoring* 11(6): 1276–1283.

Abbott, S., Caughley, B. and Douwes, J. (2007). The microbiological quality of roof-collected rainwater of private dwellings in New Zealand. In *Proceedings of the 13th International Conference on Rainwater Catchment Systems*, August 21 – 23, 2007, Sydney, Australia www.rainwater2007.com (Accessed 21-06-2023).

Abdus-Salam, N., Adekola, F.A. and Otuyo-Ibrahim (2011). Chemical composition of wet precipitation in ambient environment of Ilorin, north central Nigeria. *Journal of Saudi Chemical Society* 18(5): 528–534.

Abia, A. L. K., Ubomba-Jaswa, E., Genthe, B. and Momba, M. N. B. (2016). Quantitative microbial risk assessment (QMRA) shows increased public health risk associated with exposure to river water under conditions of riverbed sediment resuspension. *Science of the Total Environment* 566–567: 1143–1151. doi: 10.1016/j.scitotenv.2016.05.155.

Achadu, O.J., Ako, F.E. and Dalla, C.L. (2013). Quality assessment of stored harvested rainwater in Wukari, North-Eastern Nigeria: Impact of storage media. *IOSR Journal of Environmental Science, Toxicology and Food Technology (IOSR-JESTFT)* 7(5): 25–32. e-ISSN: 2319-2402, p-ISSN: 2319-2399.

Adejuwon, J.O. and Mbuk, C.J. (2011). Biological and physiochemical properties of shallow wells in Ikorodu town Lagos Nigeria. *Journal of Geology and Mining Research* 3(6): 161–168.

Adekoya, L.O. and Adewale, A.A. (1992). Wind energy potential of Nigeria. *Renewable Energy* 2(1): 35–39.

Adeniyi, I.F. and Olabanji, I.O. (2005). The physicochemical and bacteriological quality of rainwater collected over different roofing materials in Ile-Ife, south-western Nigeria. *Chemistry and Ecology* 21(3): 149–166.

Adetunji, M.T., Martins, O. and Arowolo, T.A. (2001). Seasonal variation in atmospheric deposition of nitrate, sulphate, lead, zinc, and copper in South-Western Nigeria. *Communications in Soil Science and Plant Analysis* 32(1–2): 65–73.

Afeti, G.M. and Resch, F.J. (2000). Physical characteristics of Saharan dust near the Gulf of Guinea. *Atmospheric Environment* 34(8): 1273–1279.

Ahmed, W., Gardner and Toze, S. (2011). Microbiological quality of roof-harvested rainwater and health risks: A review. *Journal of Environment Quality* 40: 1–9.

Ahmed, W., Hamilton, K., Toze, S., Cook, S. and Page, D. (2019). A review on microbial contaminants in stormwater runoff and outfalls: Potential health risks and mitigation strategies. *Science of the Total Environment* 692: 1304–1321. doi: 10. 1016/j.scitotenv.2019.07.055.

Ahmed, W., Richardson, K., Sidhu, J.P.S. and Toze, S. (2012). *Escherichia coli* and *Enterococcus spp.* in Rainwater Tank Samples: Comparison of Culture-Based Methods and 23S rRNA Gene Quantitative PCR Assays. *Environmental Science and Technology* 46(20):11370–11376. doi: 10.1021/es302222b..

Ahmed, W., Vieritz, A., Gardner, T. and Goonetilleke, A. (2009). Microbial risks from rainwater tanks in southeast Queensland. *Water* 36(8): 80–85.

Ahmed, W., Vieritz, A., Goonetilleke, A. and Gardner, T. (2010). Health risk from the use of roof-harvested rainwater in SouthEast Queensland, Australia, as potable or non-potable water, determined using Quantitative Microbial Risk Assessment. *Applied and Environment Microbiology* 76(22): 7382–7391. doi: 10.1128/AEM.00944-10.

Ajibade, F.O., Adewumi, J.R. and Oguntuase, A.M. (2014). Sustainable approach to wastewater management in the Federal University of Technology, Akure, Nigeria. *Nigerian Journal of Technological Research* 9(2): 27–36.

Akinyemi, K.O., Oyefolu, A.O., Opere, B., Otunba-Payne, V.A. and Oworu, A.O. (1998). E. coli in patients with acute gastroenteritis in Lagos, Nigeria. *East African Medical Journal* 75(9): 512–515. PMID: 10493052.

Al-Amin, M.A. (2013). Water quality study of river Kaduna in Nigeria. *International Journal of Advanced Research* 1(7): 467–474.

Al-Delaimy, A.K., Al-Meekhlafi, H.M., Nasr, N.A., Sady, H., Atroosh, W.M., Nashiry, M., Anuar, T.S., Moktar, N., Lim, Y.A.L. and Mahmud, R. (2014). Epidemiology of intestinal polyparasitism among Orang Asli school children in rural Malaysia. *PLOS Neglected Tropical Diseases* 8(8): 3074.

Al-Khashman, O.A. (2009). Chemical characteristics of rainwater collected at a western site of Jordan. *Atmospheric Research* 91(1): 53–61.

Alagbe, S.A. (2002). Groundwater resources of river Kan Gimi Basin, north-central, Nigeria. *Environmental Geology* 42(4): 404–413.

Alghamdi, M.A., Shamy, M., Redal, M.A., Khoder, M., Awad, A.H. and Elserougy, S. (2014). Microorganisms associated particulate matter: A preliminary study. *Science of the Total Environment* 479–480: 109–116. doi: 10.1016/j.scitotenv.2014.02.006. PMID: 24561289.

Amin, M.T., Kim, T.I., Amin, M.N. and Han, M.Y. (2013). Effects of catchment, first-flush, storage conditions and time on microbial quality in rainwater harvesting systems. Water Environment Research 85(12): 2317–2329.

APHA, AWWA and W.E.F. (2005). *Standard Methods for the Examination of Water and Wastewater*, 21ˢᵗ ed. American Public Health Association, World Economic Forum, American Water Works Association, Washington, DC.

Ariyananda, T. and Mawatha, E. (1999). Comparative review of drinking water quality from different rainwater harvesting systems in Sri Lanka. In *Proceedings of the 9th Conference of International Rainwater Catchment Systems Association,*

Petrolina, Brazil. International Rainwater Catchment Systems Association, pp. 1–7.

Arowolo, T.A., Taiwo, A.M., Olujimi, O.O. and Bamgbose, O. (2012). Surface water quality monitoring in Nigeria: Situational analysis and future management strategy. In: Voudouris, Kostas and Voutsa, Dimitra (Eds.), *Water Quality Monitoring and Assessment.* ISBN: 978-953-51-0486-5. OpenInTech, Rijeka.

Arsene, C., Olariu, R.I. and Mihalopoulos, N. (2007). Chemical composition of rainwater in the northeastern Romania, Iasi region (2003–2006). *Atmospheric Environment* 41(40): 9452–9467.

Ashbolt, N.J., Schoen, M.E., Soller, J.A. and Roser, D.J. (2010). Predicting pathogen risks to aid beach management: The real value of quantitative microbial risk assessment (QMRA). *Water Research* 44(16): 4692–4703.

Ashbolt, R. and Kirk, M.D. (2006). Salmonella Mississippi infections in Tasmania: The role of native Australian animals and untreated drinking water. *Epidemiology and Infection* 134(6): 1257–1265.

Asthana, D.K. and Asthana, M. (2003). *Environment: Problems and Solutions.* S.Chand and Company Ltd, New Delhi, Chapter 15, 205.

Bain, R., Cronk, R., Bonjour, S., Onda, K., Wright, J. and Yang, H. (2014). Assessment of the level of exposure to fecally contaminated drinking water in developing countries. *Tropical Medicine and International Health* 19.

Bain, R., Johnston, R. and Slaymaker, T. (2020). Drinking water quality and the SDGs. *NPJ Clean Water* 3(1): 37. 10.1038/s41545-020-00085-z.

Balderrama-Carmona, A.P., Gortáres-Moroyoqui, P., Álvarez-Valencia, L.H., Castro-Espinoza, L., Balderas-Cortés, J.J., Mondaca-Fernández, I., Chaidez-Quiroz, C. and Meza-Montenegro, M.M. (2015). Quantitative microbial risk assessment of *Cryptosporidium* and *Giardia* in well water from a native community of Mexico. *International Journal of Environmental Health Research* 25(5): 570–582.

Balogun, I. I., Sojobi, A. O., Galkaye, E. and Mannina, G. (2017). Public water supply in Lagos state, Nigeria: Review of importance and challenges, status and concerns and pragmatic solutions. *Journal of Cogent Engineering* 4(1): 1–21. doi: 10.1080/23311916.2017.1329776.

Bello, A.H. (2007a). Evaluation of Solid waste scavenging activities in Ikeja, Lagos state (Unpublished). Independent Project on Nigeria. Department of Urban and Regional Planning, Obafemi Awolowo University, Ile Ife.

Bello, A.H. (2007b). Environmental Sanitation practices in the core of Ikorodu, Lagos state. http://www.scribd.com/doc/61919266/environmental-sanitation-practices-in-the-core-of-ikorodu-lagos-state-nigeria (Accessed on June 2, 2014).

Bertrand, G., Celle-Jeanton, H., Laj, P., Rangognio, J. and Chazot, G. (2009) (2008). Rainfall chemistry: Long range transport versus below cloud scavenging. A two-year study at an inland station (Opme, France). *Journal of Atmospheric Chemistry* 60(3): 253–271).

BGS (2003). *Groundwater Quality: Nigeria.* Gritish Geological Survey. *NERC.*

Breuning-Madsen, H. and Awadzi, T.W. (2005). Harmattan dust deposition and particle size in Ghana. *CATENA* 63(1): 23–38.

Brodribb, R., Webster, P. and Farrell, D. (1995). Recurrent Campylobacter fetus subspecies fetus bacteraemia in a febrile neutropaenic patient linked to tank water. *Communicable Diseases Intelligence* 19: 312–313.

Brooks, N. and Legrand, M. (2000). Dust variability over northern Africa and rainfall in the Sahel. In: McLaren, S. and Kniveton, D. (Eds.), *Linking Climate Change to Land Surface Change*. Kluwer Academic Publishers, Dordrecht, pp. 1–25.

Cabral, C., Lucas, P. and Gordon, D. (2009). Estimating the health impacts of unsafe drinking water in developing country context. Aquatest Working Paper No. 01/09.

Caccio, S.M. (2004). New methods for the diagnosis of Cryptosporidium e-Giardia. *Parassitologia (Rome)* 46(1–2): 151–155.

Calderon, R., Mood, E. and Dufour, A. (1991). Health effects of swimmers and nonpoint sources of contaminated water. *International Journal of Environmental Health Research* 1(1): 21–31.

Carlton, E.J., Liang, S., McDowell, J.Z., Li, H.Z., Luo, W. and Remais, J.V. (2012). Regional disparities in the burden of disease attributable to unsafe water and poor sanitation in China. *Bulletin of the World Health Organization* 90(8): 578–587.

Carter, R.C. and Alhassan, A.B. (1998). Groundwater, soils, and development in the oases of the Manga grasslands, northeast Nigeria. In: Wheater, H. and Kirby, C. (Eds.), *Hydrology in a Changing Environment. Proceedings of the British Hydrological Society*. Exeter, pp. 205–211. Wiley, Chichester.

CDC (2019) Household water treatment. Global water, sanitation and hygiene (WASH). Household water treatment | Global water, sanitation and hygiene | Healthy water. CDC (Accessed on January 29, 2023).

CDC (2021). *Rainwater Collection* (Accessed on February 16, 2023). Centres for Disease Control and Prevention Rainwater Collection | Private Water Systems | Drinking Water | Healthy Water | CDC.

Chang, M., McBroom, M.W. and Beasley, R.S. (2004). Roofing as a source of nonpoint water pollution. *Journal of Environmental Management* 73(4): 307–315.

Chantara, S. and Chunsuk, N. (2008). Comparison of wet-only and bulk deposition at Chiang Mai (Thailand) based on rainwater chemical composition. *Atmospheric Environment* 42(22): 5511–5518.

Characklis, G.W., Dilts, M.J., Simmons, O.D., Likirdopulus, C.A., Krometis, L.H. and Sobsey, M.D. (2005). Microbial partitioning to settle-able particles in stormwater. *Water Research* 39(9): 1773–1782. doi: 10.1016/j.watres.2005.03.004. PMID: 15899275 24.

Chinedu, S.N., Nwinyi, O.C., Oluwadamisi, A.Y. and Eze, V.N. (2011). Assessment of water quality in Cannanland Ota, Southwest Nigeria. *Agriculture and Biology. Journal of North America* 7525: 2151.

Chukwuma, E.C., Nzediegwu, C., Umeghalu, E.C. and Ogbu, K.N. (2012). Quality Assessment of Direct Harvested Rainwater in parts of Anambra State, Nigeria. *Hydrology for Disaster Management*. Special Publication of the Nigerian Association of Hydrological Sciences.

Church, T. (2001). Rainwater harvesting: An alternative to the roof washer (Unpublished report). (web based version) (Accessed 12-09-2012).

Cinar, H.N., Gopinath, G., Jarvis, K. and Murphy, H.R. (2015). The complete mitochondrial genome of the foodborne parasitic pathogen Cyclospora cayetanensis. *PLOS ONE* 10(6): 0128645.

Clasen, T. (2006). Household water treatment for the prevention of diarrhoeal disease. PhD dissertation. University of London, London School of Hygiene & Tropical Medicine, London, UK.

Clasen, T., Pruss-Ustun, A., Mathers, C.D., Cumming, O., Cairncross and Colford, J.M. (2014). Estimating the impacts of unsafe water, sanitation and hygiene on the global burden of disease: Evolving and alternative methods. *Tropical Medicine and International Health* 90(8): 884–893.

Clasen, T.F., Thao, D.H., Boisson, S. and Shipin, O. (2008). Microbiological effectiveness and cost of boiling to disinfect drinking water in rural Vietnam. *Environmental Science and Technology* 42(12): 4255–4260.

Colford Jr., J.M., Schiff, K.C., Griffith, J.F., Yau, V., Arnold, B.F.,Wright, C.C., Gruber, J.S., Wade, T.J., Burns, S., Hayes, J., McGee, C., Gold, M., Cao, Y., Noble, R.T., Haugland, R. and Weisberg, S.B. (2012). Using rapid indicators for Enterococcus to assess the risk of illness after exposure to urban runoff contaminated marine water. *Water Research* 46(7): 2176–2186.

Coskun, O., Altayli, E., Koru, O., Tanyuksel, M. and Eyigun, C.P. (2010). Successful treatment an immune-competent patient with refractory giardiasis using nitazoxanide and genetic characterization of the Giardia intestinalis Isolate. *American Journal of Tropical Medicine and Hygiene* 83(5): 94–104.

Crabtree, K.D., Ruskin, R.H., Shaw, S.B. and Rose, J.B. (1996). The detection of Cryptosporidium oocysts and Giardia cysts in cistern water in the US Virgin Islands. *Water Research* 30(1): 208–216.

Cunliffe, D.A. (1998). Guidance on the use of rainwater tanks: National Environmental Health Forum Monographs. Water Series No. 3.

D'Almeida, G.A. (1986). A model for Saharan dust transport. *Journal of Climatology and Applied Meteorology* 25(7): 903–916.

Daso, A.P. and Osibanjo, O. (2012). Water quality issues in developing countries – A case study of Ibadan Metropolis, Nigeria. In: Dr. Voudouris (Ed.), *Water Quality Monitoring and Assessment*. Tech. ISBN: 978-953-51-0486-5. OpenInTech, Rijeka.

Davis, M. L. and Cornwell, D. A. (2008). *Introduction to Environmental Engineering*. McGraw-Hill Companies, New York.

Davies-Colley, R. J. and Smith, D. G. (2001). Turbidity suspended sediment, and water clarity: A review. *Journal of the American Water Resources Association* 37: 1085–1101. doi: 10.1111/j.1752-1688.2001.tb03624.x

Dean, J. and Hunter, P.R. (2012). Risk of gastrointestinal illness associated with the consumption of rainwater: A systematic review. *Environmental Science and Technology* 46(5): 2501–2507.

Department of Health (2022). *Rainwater* (health.vic.gov.au) (Accessed on February 16, 2023).

De Vera, G.A., Stalter, D., Gernjak, W., Weinberg, H.S., Keller, J. and Farré, M.J. (2015). Towards reducing DBP formation potential of drinking water by favouring direct ozone over hydroxyl radical reactions during ozonation. *Water Research* 87: 49–58.

DFID (2015). 2010 to 2015 government policy: Water and sanitation in developing countries. Department for International Development. https://www.gov.uk/government/publications/2010-to-2015-government-policy-water-and-sanitation-in-developing-countries/2010-to-2015-government-policy-water-and-sanitation-in-developing-countries (Accessed on October 20, 2015).

Dolske, D.A. and Gatz, D.F. (1985). A field intercomparison of methods for the measurement of particle and gas dry deposition. *Journal of Geophysical Research, Atmospheres* 90(D1): 2076–2084.

Doyle, K. C. (2006). Sizing the first flush and its effect on the storage reliability yield behaviour of rainwater harvesting in Rwanda. Published Thesis, includes bibliographical references, 112–119. http://hdl.handle.net/1721.1/44289.

Draper, N. R. and Smith, H. (1981). *Applied Regression Analysis*, 2nd Edition, John Wiley & Sons, New York.

Efe, S.I. (2006). Quality of rainwater harvesting for rural communities of Delta State, Nigeria. *Environmentalist* 26(3): 175–181.

Efe, S.I. (2010). Spatial variation in acid and some heavy metal composition of rainwater harvesting in the oil-producing region of Nigeria. *Natural Hazards* 55(2): 307–319.

Eletta, O.A.A. and Oyeyipo, J.O. (2008). Rainwater harvesting: Effect of age of roof on water quality. *International Journal of Applied Chemistry* 2: 157–162. ISSN 0973-17924.

Embry, S. S. (2001). Microbiological quality of puget sound basin streams and identification of contaminant sources. *Journal of the American Water Resources Association* 37(2): 407–421.

EnHealth (2004). *EnHealth*. Subcommittee of National Public Health Partnership of the Australian Government.

EPA (2001). *Frequently Asked Questions about Atmospheric Deposition: A Handbook for Watershed Managers*. Office of Wetlands, Oceans, and Watersheds. U.S. Environmental Protection Agency, Washington, DC, p. 20460.

Eruola, A.O., Ufoegbune, G.C., Eruola, A.O., Ojekunle, Z.O., Makinde, A.A. and Amori, A.A. (2012). Qualitative assessment of the effect of thunderstorm on rainwater harvesting from rooftop catchments at Oke-Lantoro Community in Abeokuta, Southwest Nigeria. *Resources and Environment* 2(1): 27–32.

EU WFD (2000). Water framework directive (WFD) 2000/60/EC: Directive 2000/60/EC of the European parliament and of the council of 23 October 2000 establishing a framework for Community action in the field of water policy. Water Framework Directive (WFD) 2000/60/EC — European Environment Agency (europa.eu) (Accessed on May 30, 2023).

Fagbenle, R.L. and Karayiannis, T.G. (1994). On the wind energy resource of Nigeria. *International Journal of Energy Research* 18(5): 493–508.

Ferguson, R. I. (1986). River loads underestimated by rating curves. *Water Resources Research* 22(1): 7476.

FGN (2000). Water Supply & Interim Strategy note. Federal Government of Nigeria. Available at:http://siteresources.worldbank.org/NIGERIAEXTN/Resources/wss_1100.pdf (Accessed on June 17, 2010).

Forster, J. (1996). Patterns of roof runoff contamination and their potential implications on practice and regulations of treatment and local infiltration. *Water Science and Technology* 33(6): 39–48.

Freeflush (2017) First Flush Filters and Diverters, Theory, and Application. First Flush Filters and Diverters, Theory, and Application. Freeflush Water Management Ltd. (Accessed on February 8, 2023).

Georgakopoulos, D.G., Despres, V., Frohlich-Nowoisky, J., Psenner, R., Ariya, P.A., Posfai, M., Ahern, H.E., Moffett, B.F. and Hill, T.C.J. (2009). Microbiology and atmospheric processes: Biological, physical, and chemical characterization of aerosol particles. *Biogeosciences* 6(4): 721–737.

George, J., An, W., Joshi, D., Zhang, D.Q., Yang, M. and Suriyanarayanan, S. (2015). Quantitative microbial risk assessment to estimate the health risk in urban drinking water systems of Mysore, Karnataka, India. *Water Quality, Exposure and Health* 7(3): 331–338.

Ghisi, E. and Ferreira, D.F. (2007). Potential for potable water savings by using rainwater and greywater in a multi-storey residential building in southern Brazil. *Building and Environment* 42(7): 2512–2522.

Gikas, G.D. and Tsihrintzis, V.A. (2012). Assessment of water quality of first-flush roof runoff and harvested rainwater. Journal of Hydrology 466–467: 115–126.

Goncalves, F.L.T., Martins, J.A., Albrecht, R.I., Morales, C.A., Silva Dias, M.A. and Morris, C.E. (2012). Effect of bacterial ice nuclei on the frequency and intensity of lightning activity inferred by the BRAMS model. *Atmospheric Chemistry and Physics* 12(13): 5677–5689.

Good, J.C. (1993). Roof runoff as a diffuse source of metals and aquatic toxicity in storm water. *Water Science and Technology* 28(3–5): 317–321.

Google Earth (2020). Google Earth pictures of Ikorodu Area of Lagos State, Nigeria. Available from: https://earth.google.com/ web/@6.60511854,3.52450822,16. 6434089a,23128.70908794d,35y,0h,0t,0r/data=CjgaNhIwCiUweDEwM2JlZT YyZDQ0YWI1NzM6MHg0NGRkZjZlYTE4ZWRmmY2IzKgdJa29yb2R1GAIgA Q (Accessed on September 1, 2020).

Gould, J. (1999). Is rainwater safe to drink? A review of recent findings. Proceedings of the 9th International Rainwater Catchment Systems Conference, 1999. http://www.eng.warwick.ac.uk/ircsa/abs/9th/7_4.html (Accessed on March 13, 2011).

Grantz, D.A., Garner, J.H.B. and Johnson, D.W. (2003). Ecological effects of particulate matter. *Environment International* 29(2–3): 213–239. doi: 10.1016/S0160-4120(02)00181-2.

Gray, N.F. (2010). *Water Technology: An Introduction for Environmental Scientists and Engineers*, 3rd ed. ISBN-13, p. 978-1856177054. CRC Press, London.

Gregory, R. and Edzwald, J. (2010). Chapt.9. Sedimentation & flotation. In: *Water Quality & Treatment*, 6th ed. AWWA & McGrawHill, Denver, Colorado,.

Guerzoni, S., Chester, R., Dulac, F., Herut, B., Loye-Pilot, M.D., Measures, C., Migon, C., Molinaroli, E., Moulin, C., Rossini, P., Saydam, C., Soudine, A. and Ziveri, P. (1999). The role of atmospheric deposition in the biogeochemistry of the Mediterranean Sea. *Progress in Oceanography* 44(1–3): 147–190.

Gunawardana, C., Goonetilleke, A., Egodawatta, P., Dawes, L. and Kokot, S. (2012). Source characterisation of road dust based on chemical and mineralogical composition. *Chemosphere* 87(2): 163–170.

Haas, C.N., Rose, J.B. and Gerba, C.P. (1999). *Quantitative Microbial Risk Assessment*. John Wiley & Sons, Inc., New York, p. 449.

Haile, R.W., Witte, J.S., Gold, M., Cressey, R., McGee, C., Millikan, R.C., Glasser, A., Harawa, N., Ervin, C., Harmon, P., Harper, J., Dermand, J., Alamillo, J., Barrett, K., Nides, M. and Wang, G.Y. (1999). The health effects of swimming in ocean water contaminated by storm drain runoff. *Epidemiology* 10(4): 355–363.

Hamilton, K.A., Ahmed, W., Toze, S. and Haas, C.N. (2017). Human health risks for Legionella and Mycobacterium avium complex (MAC) from potable and non-potable uses of roof-harvested rainwater. *Water Research* 119: 288–303.

Handia, L. (2005). Comparative study of rainwater quality in urban Zambia. *Journal of Water Supply Research and Technology – Aqua* 54(1): 55–64.

Handidu, J.A. (1990). National growth, water demand and supply strategies in Nigeria in the 1960s. *Journal of the Nigerian Association of Hydrogeologists (NAH)* 2(1): 35–44.

Hartung, H. (2007). Rainwater utilization in Africa, some experiences. In: *Proceedings of the 1ˢᵗ International Rainwater Leadership Workshop*. International Water Association. Rainwater Harvesting and Management Specialist Group. Seoul National University Brain Korea 21 Sustainable Infrastructure Research Group (SNU BK21 SIR Group). United Nations Environment Programme (UNEP), Seoul, Korea (IWA. WHM SG).

Hathaway, J.M. and Hunt, W.F. (2011). Evaluation of first flush for indicator bacteria and total suspended solids in urban stormwater runoff. *Water Air Soil Pollution Journal* 217(1–4): 135–147. 10.1007/s11270-010-0574-y.

Hatibu, N., Mutabazi, K., Senkondo, E.M. and Msangi, A.S.K. (2006). Economics of rainwater harvesting for crop enterprises in semi-arid areas of East Africa. *Agricultural Water Management* 80(1–3): 74–86.

Helmer, R. and Hesanhol, I. (1997). *Water Pollution Control: A Guide to the Use of Water Quality Management Principles*. The United Nations Environment Programme, the Water Supply & Sanitation Collaborative Council and the World Health Organization. E. & F. Spon.

Herrador, B.R.G., Blasio, B.F., MacDonald, E., Nichols, G., Sudre, B., Vold, L., Semenza, J.C. and Nygård, K. (2015). Analytical studies assessing the association between extreme precipitation or temperature and drinking water-related waterborne infections: A review. *Environmental Health*: 14–29.

Hill, R. (2006). Bacterial activity in harvested rainwater. *Whitewater Ltd Consulting Engineers and Scientist*, 1–8, http://www.whitewaterlimited.com/BacterialActivi tyinHarvestedRainWater.pdf.

Hora, J., Cohim, E. B., Sipert, S. and Leão, A. (2018). Quantitative microbial risk assessment (QMRA) of campylobacter for roof-harvested rainwater domestic use. *Proceedings of MDPI* 2(5): 185–194. doi: 10.3390/ecws-2-04954.

Howard, G. and Bartram, J. (2003). *Domestic Water Quantity, Service Level, and Health* (p. 33). World Health Organization, Geneva. http://www.ircwash.org/ resources/domestic-water-quantity-service-level-and-health (Accessed on October 7, 2015).

Huey, G. M. and Meyer, M. L. (2010). Turbidity as an indicator of water quality in diverse watersheds of the Upper Pecos River Basin. *Water* 2(2): 273–284.

Hunter, P.R., Payment, P., Ashbolt, N. and Bartram, J. (2003) Chapter 3. Assessment of risk. In: Ronchi, E. and Bartram, J. (Eds.), *Assessing Microbial Safety of Drinking Water: Improving Approaches and Methods*. OECD/WHO, Paris Guidance document. OECD/WHO, pp. 79–109.

Hussain, S., Leeuwen, J.V., Chow, C.W.K., Aryal, R., Beecham, S., Duan, J. and Drikas, M. (2014). Comparison of the coagulation performance of tetravalent titanium and zirconium salts with alum. *Chemical Engineering Journal* 254: 635–646.

IDEXX (2023). Colilert - IDEXX US (Accessed on January 14, 2023).

IDEXX Laboratories (2016). A new standard in coliform/E.coli detection. https:// www.idexx.com/water/products/colilert-18.html (Assessed on April 29, 2016).

Ikelionwu, C. (2006). *Water Quality Monitoring and Surveillance in Nigeria*. Department of Water Supply and Quality Control, Federal Ministry of Water

Resources,   Abuja   http://nwri.gov.ng/userfiles/file/FMWR_Water_Quality_ Surveillence_draft.pdf (Accessed on October 2, 2014).

Irvine, K. N., Somogye, E. L. and Pettibone, G. W. (2002). Turbidity, suspended solids, and bacteria relationships in the Buffalo River Watershed. *Middle States Geographer* 35: 42–51.

Islam, M.A., Sakakibara, H., Karim, M.R., Sekine, M. and Mahmud, Z.H. (2011). Bacteriological assessment of drinking water supply options in coastal areas of Bangladesh. *Journal of Water and Health* 9(2): 415–428.

ISO-17994 (2004). *Water Quality Criteria for Establishing Equivalence between Microbiological Methods*. International organization for Standardization (ISO), Geneva, Switzerland.

ISO-1908-1 (2000). *Water Quality: Detection and Enumeration of Escherichia coli and Coliform Bacteria Part 1: Membrane Filtration Method*. International Organization for Standardization (ISO), Geneva, Switzerland, p. 10.

Izquierdo, R. and Avila, A. (2012). Comparison of collection methods to determine atmospheric deposition in a rural Mediterranean site (NE Spain). *Journal of Atmospheric Chemistry* 69(4): 351–368.

John, C.K., Pu, J.H., Pandey, M. and Hanmaiahgari, P.R. (2021a). Sediment deposition within rainwater: Case study comparison of four different sites in Ikorodu, Nigeria. *Fluids* 6(3), 124: 2021. doi: 10.3390/ fluids6030124.

John, C.K., Pu, J.H., Pandey, M. and Moruzzi, R. (2021b). Impacts of sedimentation on rainwater quality: Case study at Ikorodu of Lagos, Nigeria. *Water Supply*. doi: 10.2166/ws.2021.093.

John, C.K., Pu, J.H., Moruzzi, R., Pandey, M. and Azamathulla, H. (2021c). Reusable rainwater quality at Ikorodu area of Lagos, Nigeria: Impact of first-flush and household treatment techniques. *Journal of Water, Sanitation and Hygiene for Development*. doi: 10.2166/washdev.2021.062.

John, C.K., Pu, J.H., Moruzzi, R. and Pandey, M. (2021d). Health-risk assessment for roof-harvested rainwater via QMRA in Ikorodu area, Lagos, Nigeria. *Journal of Water and Climate Change*. doi: 10.2166/wcc.2021.025.

Johnson, M., Iwalokun, B.A., Longe, O.A. and Babalola, O.M. (2014). Antimicrobial susceptibility patterns of methanolic leaf extract of Azadirachta indica and some selected antibiotics and plasmid profiles of Escherichia coli isolates obtained from different human clinical specimens in Lagos- Nigeria. *Science Journal of Microbiology*. ISSN: 2276-6359. doi: 10.7237/sjmb/152.

Jones, M.P. and Hunt, W.F. (2008). Rainwater harvesting: Guidance for homeowners. Available at http://www.ces.ncsu.edu/depts/agecon/WECO/documents/WaterHa rvestHome2008.pdf (Accessed on January 22, 2013).

Julian, T. R., Islam, M. A., Pickering, A. J., Roy, S., Fuhrmeister, E. R., Ercumen, R., Harris, A., Bishai, J. and Schwab, K. J. (2015). Genotypic and phenotypic characterization of Escherichia coli isolates from feces, hands, and soils in rural Bangladesh via the colilert quanti-tray system. *Applied and Environmental Microbiology* 81(5), 1735–1743. doi: 10.1128/AEM.03214-14.

Junaid, S.A. and Agina, S.E. (2014). Sanitary survey of drinking water quality in Plateau State, Nigeria. *British Biotechnology Journal* 4(12): 1313.

Kaldellis, J.K. and Kondili, E.M. (2007). The Water shortage problem in the Aegean archipelago islands: Cost-effective desalination prospects. *Desalination* 216(1–3): 123–138.

Kanarat, S. (2004). In: Cotruvo, J.A., Dufour, A., Rees, J., Bartram, J., Carr, R., Cliver, D.O., Craun, G.F., Fayer, R. and Gannon, V.P.J. (Eds.), *Waterborne Zoonoses: Identification, Causes and Control*. IWA Publishing, London, UK, pp. 136–150.

Kaushal, S.S., Likens, G.E., Pace, M.L., Utz, R.M., Haq, S., Gorman, J. and Grese, M. (2018). Freshwater salinization syndrome on a continental scale. *Proceedings of the National Academy of Sciences of the United States of America* 115(4): 574–583.

KDHS (2014). Kenya demographic and health survey. Kenya - Kenya Demographic and Health Survey 2014 (knbs.or.ke) (Accessed on May 30, 2023).

Koc-ak, M., Kubilay, N. and Mihalopoulos, N. (2004a). Ionic composition of lower tropospheric aerosols at a Northeastern Mediterranean site: Implications. Regarding sources and long-range transport. *Atmospheric Environment* 38: 142067–142077.

Koc-ak, M., Nimmo, M., Kubilay, N. and Herut, B. (2004b). Spatio-temporal aerosol trace metal concentrations and sources in the Levantine basin of the Eastern Mediterranean. *Atmospheric Environment* 38(14): 2133–2144.

Krometis, L.A., Noble, R.T., Characklis, G.W., Blackwood, D. and Sobsey, M.D. (2013). Assessment of *E. coli* partitioning behaviour via both culture-based and qPCR methods. *Water Science and Technology* 68(6): 1359–1369. doi: 10.2166/wst.2013.363.

Kuruk, P. (2005). Customary water laws and practices: Nigeria. Available at http://www.fao.org/Legal/advserv/FAOIUCNcs/Nigeria.pdf. (Accessed on June 17, 2010).

Kus, B., Kandasamy, J., Vigneswaran, S. and Shon, H.K. (2010). Analysis of first flush to improve the water quality in rainwater tanks. *Water Science and Technology* 61(2): 421–428.

Lakshminarayanan, S. and Jayalakshmy, R. (2015). Diarrheal diseases among children in India: Current scenario and future perspectives. *Journal of Natural Science, Biology, and Medicine* 6: 24–28. doi: 10.4103/0976-9668.149073

Lammerding, A.M. and Fazil, A. (2000). Hazard Identification and exposure assessment for microbial food safety risk assessment. *International Journal of Food Microbiology* 58(3): 147–157.

Lawrence, S. J. (2012). *Escherichia Coli Bacteria Density in Relation to Turbidity, Streamflow Characteristics, and Season in the Chattahoochee River Near Atlanta, Georgia, October 2000 Through September 2008—Description, Statistical Analysis, and Predictive Modeling: U.S. Geological Survey Scientific Investigations Report 2012–5037*, p. 81.

Lee, J.Y., Bak, G. and Han, M. (2012). Quality of roof-harvested rainwater – Comparison of different roofing materials. *Environmental Pollution* 162: 422–429.

Lee, J.Y., Yang, J.S., Han, M. and Choi, J. (2010). Comparison of the microbiological and chemical characterization of harvested rainwater and reservoir water as alternative water resources. *Science of the Total Environment* 408(4): 896–905.

Leite, D. B. P. & Moruzzi, R. B. (2016). Considerations on the maximum permitted values (VMP) of E. coli for grey water reuse by means of quantitative risk assessment of microbiology (QRMA). *Engenharia Sanitaria E Ambiental* 22(1): 57–64. doi: 10.1590/s1413-41522016119617.

Lester, R. (1992). A mixed outbreak of cryptosporidium and giardiasis. *Update, Australia* 1(1): 14–15.

Lim, K.Y. and Jiang, S.C. (2013). Re-evaluation of health risk benchmark for sustainable water practice through risk analysis of rooftop-harvested rainwater. *Water Research* 47(20): 7273–7286.

Lim, S.S., Vos, T. and Flaxman, A.D. (2012). A comparative risk assessment of burden of disease and injury attributable to 67 risk factors and risk factor clusters in 21 regions, 1990–2010: A systematic analysis for the Global Burden of Disease Study. *Lancet* 380(9859): 2224–2260.

Longe, E.O., Omole, D.O., Adewumi, I.K. and Ogbiye, A.S. (2010). Water resources use, abuse and regulations in Nigeria. *Journal of Sustainable Development in Africa* 12(2): 1–10.

Lovett, G.M. (1994). Atmospheric deposition of nutrients and pollutants in North America: *An ecological perspective. Ecological Applications* 4(4): 629–650.

Lozano, R. et al. (2012). Global and regional mortality from 235 causes of death for 20 age groups in 1990 and 2010: A systematic analysis for the Global Burden of Disease Study 2010. *Lancet* 380(9859): 2095–2128.

Lye, D.J. (2002). Health risks associated with consumption of untreated water from household roof catchment systems 1. *JAWRA Journal of the American Water Resources Association* 38(5): 1301–1306.

Lye, D. J. (2002). Health risks associated with consumption of untreated water from household roof catchment systems. *Journal of the American Water Resources Association* 38(5): 1301–1306. doi: 10.1111/j.1752-1688.2002.tb04349.x.

Lye, D. J. (2009). Rooftop Runoff as a Source of contamination: A review. *Science of the Total Environment* 407: 5429–5434. doi: 10.1016/j.scitotenv.2009.07.011.

Lyngsie, G., Awadzi, T. and Breuning-Madsen, H. (2011). Origin of Harmattan dust settled in Northern Ghana – Long transported or local dust? *Geoderma* 167–168: 351–359.

Mahmood, K. (1987). Reservoir sedimentation: Impact, extent, and mitigation (Technical paper). http://www.osti.gov/scitech/biblio/5564758 (Accessed on December 23, 2015).

Machdar, E., Van der Steen, N. P., Raschid-Sally, L. & Lens, P. N. L. (2013). Application of quantitative microbial risk assessment to analyze the public health risk from poor drinking water quality in a low-income area in Accra, Ghana. *Science of the Total Environment* 449: 134–142. doi: 10.1016/j.scitotenv.2013.01.048.

Mannapperuma, W.M.G.C.K., Abayasekara, C.L., Herath, G.B.B., Werellagama, D.R.I.B. and Heinonen-Tanski, H. (2011). Comparison of bacteriological methods for detecting and enumerating total coliforms and *E. coli* in water. *Research Journal of Microbiology* 6(12): 851–861.

Manetu, W., M'masi, S. and Recha, C. (2021). Diarrhea disease among children under 5 years of age: A global systematic review. *Open Journal of Epidemiology* 11(3): 207–221. doi: 10.4236/ojepi.2021.113018.

Maplandia (2015). Google Maps world gazetteer. Malandia.com.Nigeria/Lagos/iko rodu/ikorodu (Accessed on October 2, 2015).

Martin, J.H., Gordon, R.M. and Fitzwater, S.E. (1991). The case for iron. *Limnology and Oceanography* 36(8): 1793–1802.

Martin, E. C. and Gentry, T. J. (2016). Impact of enumeration method on diversity of Escherichia coli genotypes isolated from surface water. *Letters in Applied Microbiology* 63(5): 369–375. doi: 10.1111/lam.12633.

Martinson, D.B. (2007). Improving the viability of roofwater harvesting in low-income countries. Unpublished PhD thesis. University of Warwick, School of Engineering.

Marcazzan, G.M. and Persico, F. (1996) (Suppl. 1). Evaluation of mixing layer depth in Milan Town from temporal variations of atmospheric radioactive aerosols. *Journal of Aerosol Science* 27: S11–S22.

Martinson, D.B. and Thomas, T. (2005). Quantifying the first-flush phenomenon. In *12th International Rainwater Catchment Systems Conference*, New Delhi, India.

Matondo, J.I., Peter, G. and Msibi, K.M. (2005). Managing water under climate change for peace and prosperity in Swaziland. *Physics and Chemistry of the Earth, Parts A/B/C* 30(11–16): 943–949.

McTainsh, G. (1980). Harmattan dust deposition in northern Nigeria. *Nature* 286(5773): 587–588.

Meera, V. and Ahammed, M.M. (2006). Water quality of rooftop rainwater harvesting systems: A review. *Journal of Water Supply Research and Technology – Aqua* 55(4): 257–268.

Mendez, C.B., Klenzendorf, J.B., Afshar, B.R., Simmons, M.T., Barrett, M.E., Kinney, K.A. and Jo Kirisits, M. (2011). The effect of roofing material on the quality of harvested rainwater. *Water Research* 45(5): 2049–2059.

Merritt, A., Miles, R. and Bates, J. (1999). An outbreak of Campylobacter enteritis on an island resort, north Queensland. *Communicable Diseases Intelligence* 23(8): 215–219.

Michaelides, G. (1987). Laboratory experiments on efficiency of foul flush diversion systems. In *3rd Annual International Rainwater Cistern Systems Conference*.

Moberg, J.P., Esu, I.E. and Malgwi, W.B. (1991). Characteristics and constituent compositioin of Harmattan dust falling in Northern Nigeria. *Geoderma* 48(1–2): 73–81.

ModernDevice (2017). Search: 12 Results Found for "Product Code MD0550" – Modern Device (Accessed on January 15, 2018).

Morakinyo, O.M., Mokgobu, M.I., Mukhola, M.S. and Godobedzha, T. (2019). Biological composition of respirable particulate matter in an industrial vicinity in South Africa. *International Journal of Environmental Research and Public Health* 21(4): 629. doi: 10.3390/ijerph16040629. PMID: 30795513, PMCID: PMC6406656.

Morris, C.E., Sands, D.C., Bardin, M., Jaenicke, R., Vogel, B., Leyronas, C., Ariya, P.A. and Psenner, R. (2011). Microbiology and atmospheric processes: Research challenges concerning the impact of airborne micro-organisms on the atmosphere and climate. *Biogeosciences* 8(1): 17–25.

Moruzzi, R. B. and Nakada, L. Y. K. (2015). Corn starch-based treatment improves rainwater quality. *Water Supply* 15(6): 1326–1333. doi: 10.2166/ws.2015.097.

Mulitza, S., Heslop, D., Pittauerova, D., Fischer, H.W., Meyer, I., Stuut, J.B., Zabel, M., Mollenhauer, G., Collins, J.A., Kuhnert, H. and Schulz, M. (2010). Increase in African dust flux at the onset of commercial agriculture in the Sahel region. *Nature* 466(7303): 226–228.

Nsi, E.W. (2007). *Basic Environmental Chemistry*, 8. The Return Press Ltd, Makurdi, p. 87.

Ntale, H.K. and Moses, N. (2003). Improving the quality of harvested rainwater by using first flush interceptors/retainers. In: *11th International Rainwater Catchment Systems Conference*. Texcoco, Mexico.

NWRI (2012). National water Research Institute: Water resources data management system. http://nwri.gov.ng/about/aboutus.asp?Index=154 (Accessed on December 12, 2012).

Okpoko, E., Egboka, B., Anike, L. and Okoro, E. (2013). Rainfall harvesting as an alternative water supply in water stressed communities in Aguata-Awka area of Southeastern Nigeria. *Environmental Engineering Research* 18(2): 95–101.

Olabanji, I.O. and Adeniyi, I.F. (2005). Trace metals in bulk freefall and roof intercepted rainwater at Ile-Ife, Southwest Nigeria. *Chemistry and Ecology* 21(3): 167–179.

Olaleye, O.N. (2012). Detection of faecal coliforms in biofilms of water tankers in Odongunyan, Ikorodu-A Peri-Urban Lagos settlement. *COLERM Proceedings* 2: 458–463.

Olaoye, R.A. and Olaniyan, O.S. (2012). Quality of rainwater from different roof material. *International Journal of Engineering and Technology* 2(8): 1413–1421.

Olobaniyi, S.B. and Efe, S.I. (2007). Comparative assessment of rainwater and groundwater quality in an oil producing area of Nigeria: Environmental and health implications. *JEHR* 6(2): 111–118.

Olowoyo, D.N. (2011). Physicochemical characteristics of rainwater quality of Warri axis of Delta state in Western Niger Delta region of Nigeria. *Journal of Environmental Chemistry and Ecotoxicology* 3(12): 320–322.

Olson, M. E., Cer, H., Morck, D. W., Buret, A. G. and Read, R. R. (2002). Biofilm bacteria: Formation and comparative susceptibility to antibiotics. *Canadian Journal of Veterinary Research* 66(2): 86–92.

Omega (2017). Omega OM-CP-WIND101A-KIT series. Cup Anemometer Wind Speed Data Logger – Order (omega.co.uk) (Accessed on December 16, 2017).

Omelia, C. (1998). Coagulation and sedimentation in lakes, reservoirs and water treatment plants. *Water Science and Technology* 37(2): 129. doi: 10.1016/S0273-1223(98)00018-3.

Onda, K., LoBuglio, J. and Bartram, J. (2012). Global access to safe water: accounting for water quality and the resulting impact on MDG progress. *International Journal of Environmental Research and Public Health* 9(3): 880–894. doi: 10.3390/ijerph9030880.

Pacey, A. and Cullis, A. (1986). *Rainwater Harvesting: The Collection of Rainfall and Runoff in Rural Areas*. AT Microfiche Reference Library.

Painter, J.E., Gargano, J.W., Collier, S.A. and Yoder, J.S. (2015). Giardiasis surveillance – United States, 2011–2012. *MMWR Surveillance Summaries* 64(3): 15–25.

Pan, Y.P. and Wang, Y.S. (2015). Atmospheric wet and dry deposition of trace elements at 10 sites in Northern China. *Atmospheric Chemistry and Physics* 15(2): 951–972.

Paton, E. and Haacke, N. (2021). Merging patterns and processes of diffuse pollution in urban watersheds: A connectivity assessment. *Wires Water* 8(4): e1525.

Payment, P., Siemiatycki, J., Richardson, L., Renaud, G., Franco, E. and Prevost, M. (1999). A prospective epidemiological study of gastrointestinal health effects due to the consumption of drinking water. *International Journal of Environmental Health Research* 7(1): 5–31.

People for rainwater Japan (2003). *Promoting Rainwater Utilization.* Promoting Rainwater Utilization (mobile)l Japan for Sustainability (japanfs.org) (Accessed on February 8, 2023).

Petterson, S. R., Mitchell, V. G., Davies, C. M., O'Connor, J., Kaucner, C., Roser, D. and Ashbolt, N. (2010). Evaluation of three full-scale stormwater treatment systems with respect to water yield, pathogen removal efficacy and human health risk from faecal pathogens. *Science of the Total Environment* 543: 691–702. doi: 10.1016/j.scitotenv.2015.11.056.

Petterson, S.R., Signor, R.S. and Ashbolt, N.J. (2007). Incorporating method recovery uncertainties in stochastic estimates of raw water protozoan concentrations for QMRA. *Journal of Water and Health* 5 Supplement 1: 50–65.

Pinfold, J.V., Horan, N.J., Wiroganagud, W. and Mara, D. (1993). The bacteriological quality of rainjar water in rural northeast Thailand. *Water Research* 27(2): 297–302.

Pinker, R.T., Pandithurai, G., Holben, B.N., Dubovik, O. and Aro, T.O. (2001). A dust outbreak episode in sub-Sahel West Africa. *Journal of Geophysical Research: Atmospheres* 106(D19): 22923–22930.

Poudyal, S., Chochrane, T.A. and Bell-Mendoza, R. (2016). First flush stormwater pollutants from carparks in different urban settings. Water New Zealand Conference Paper Nov./Dec., pp. 24–27.

Poudyal, S., Cochrane, T.A. and Bello-Mendoza, R. (2021). Carpark pollutant yields from first flush stormwater runoff. *Environmental Challenges* 5: 100299. doi: 10.1016/j.envc.2021.100301.

Pouleur, S., Richard, C., Martin, J.G. and Antoun, H. (1992). Ice nucleation activity in *Fusarium acuminatum* and *Fusarium avenaceum. Applied and Environment Microbiology* 58(9): 2960–2964.

Prathibha, P., Kothai, P., Sardhi, I.V., Pandit, G.G. and Puranik, V.D. (2010). Chemical characterization of precipitation at a coastal site in Trombay, Mumbai. India. *Environmental Monitoring and Assessment* 168(1–4): 45–53.

Prospero, J.M. (1999). Long-term measurements of the transport of African mineral dust to the southern United States: Implications for regional air quality. *Journal of Geophysical Research* 104(13): 15917–15927.

Prospero, J.M. (2011). Long term trends in African dust transport to the Caribbean: African sources, changing Climate, and future scenarios. In: *First International Workshop on the Long-Range Transport and Impacts of African Dust on the Americas*, 6–7 October.

Prospero, J.M. and Lamb, P.J. (2003). African droughts and dust transport to the Caribbean: Climate change implications. *Science* 302(5647): 1024–1027.

Prospero, J.M. and Arimoto, R. (2009). Atmospheric transport and deposition of particulate material to the oceans. *Encyclopedia of Oceanic Sciences*. doi: 10.1016/B978-012374473-9.00643-3.

Prospero, J.M., Ginoux, P., Torres, O., Nicholson, S.E. and Gill, T.E. (2002). Environmental characterization of global sources of atmospheric soil dust identified with the Nimbus 7 total ozone mapping spectrometer (TOMS) absorbing aerosol product. *Reviews of Geophysics* 40(1): 1002.

Pruppacher, H.R. and Klett, J.D. (1997). Microphysics of clouds and precipitation, Atmospheric and oceanographic sciences library. *EDN*. Kluwer, Dordrecht, 2. Rev. and Enl, 18.

Pruss, A., Kay, D., Fewtrell, L. and Bartram, J. (2002). Estimating the burden of disease from water, sanitation and hygiene at a global level. *Environmental Health Perspectives* 110(5): 537–542. doi: 10.1289/ehp.110-1240845.

Psutka, R., Peletz, R., Michelo, S., Kelly, K. and Clasen, T. (2011). Assessing the microbiological performance and potential cost of boiling drinking water in urban Zambia. *Environmental Science and Technology* 45(14): 6095–6101.

Pu, J.H. (2016). Conceptual hydrodynamic-thermal mapping modelling for coral reefs at South Singapore sea. *Applied Ocean Research* 55: 59–65.

Pu, J.H. (2021). Velocity profile and turbulence structure measurement corrections for sediment transport-induced water-Worked bed. *Fluids* 6(2): 86.

Pu, J.H., Hussain, K., Shao, S. and Huang, Y. (2014). Shallow sediment transport flow computation using time-varying sediment adaptation length. *International Journal of Sediment Research* 29(2): 171–183.

Pu, J.H., Huang, Y., Shao, S. and Hussain, K. (2016). Three-gorges dam fine sediment pollutant transport: Turbulence SPH model simulation of multi-fluid flows. *Journal of Applied Fluid Mechanics* 9(1): 1–10.

Pu, J.H. and Lim, S.Y. (2014). Efficient numerical computation and experimental study of temporally long equilibrium scour development around abutment. *Environmental Fluid Mechanics* 14(1): 69–86.

Pushpangadan, K., Sivanandan, P.K. and Joy, A. (2001). Comparative analysis of water quality from DRWH and traditional sources – A case study in Kerala, India. In: *Proceedings of the Rainwater Harvesting Conference*, New Delhi, India, pp. E4.1–E4.7.

Qiao, X., Tang, Y., Hu, J., Zhang, S., Li, J., Wu, L., Gao, H., Zhang, H., Ying, Q. and Ying, Q. (2015a). Modelling dry and wet deposition of sulfate, nitrate, and ammonium ions in Jiuzhaigou National Nature Reserve, China using a source oriented CMAQ model: Part I. Base case model results. Science of the Total Environment 532: 831–839.

Qiao, X., Tang, Y., Hu, J., Zhang, S., Li, J., Wu, L., Gao, H., Zhang, H. and Ying, Q. (2015b). Modelling dry and wet deposition of sulfate, nitrate, and ammonium ions in Jiuzhaigou National Nature Reserve, China using a source oriented CMAQ model: Part II. Emission sector and source region contributions. Science of the Total Environment 532: 840–848.

Reiner, R.C., et al. (2018). Variation in childhood diarrheal morbidity and mortality in Africa, 2000–2015. *The New England Journal of Medicine* 379(12): 1128–1138.

Resch, F., Sunnu, A. and Afeti, G. (2007). Saharan dust flux and deposition rate near the Gulf of Guinea. *Tellus B* 60B(1): 98–105.

Revitt, D.M., Lundy, L., Coulon, F. and Fairley, M. (2014). The sources, impact, and management of car park runoff pollution: A review. Journal of Environmental Management 146: 552–567. ISSN 0301-4797. doi: 10.1016/j. jenvman.2014.05.041.

Rodriguez-Alvarez, M.A., Weir, M.H., Pope, J.M., Seghezzo, L., Rajal, V.B., Salusso, M.M. and Morana, L.B. (2015). Development of a relative risk model for drinking water regulation and design recommendations for a peri-urban region of Argentina. International Journal of Hygiene and Environmental Health 218(7): 627–638.

Roser, D.J., Davies, C.M., Ashbolt, N.J. and Morison, P. (2006). Microbial exposure assessment of an urban recreational lake: A case study of the application of new risk-based guidelines. *Water Science and Technology* 54(3): 245–252.

Savou (2021). An exploration of the first flush diverter: To first flush, or not to flush. BlueBarrel rainwater catchment systems. First Flush Diverter: To Use One or Not? We Ask a Rainwater Professional! (bluebarrelsystems.com) (Accessed on February 14, 2023).

Sazakli, E., Alexopoulos, A. and Leotsinidis, M. (2007). Rainwater harvesting, quality assessment and utilization in Kefalonia Island, Greece. *Water Research* 41(9): 2039–2047.

Schets, F.M., Italiaander, R., van den Berg, H.H.J.L. and Husman, A.M.D. (2010). Rainwater harvesting; quality assessment and utilization in the Netherlands. *Journal of Water and Health* 8(2): 224–235.

Seinfeld, J.A. and Pandis, S.N. (1998). *Atmospheric Chemistry and Physics: From Air Pollution to Climate Change*. John Wiley and Sons, Inc, New York

Sharma, S. and Bhattacharya, A. (2017). Drinking water contamination and treatment techniques. *Applied Water Science* 7(3): 1043–1067.

Simmons, G., Hope, V., Lewis, G., Whitmore, J. and Gao, W. (2001). Contamination of potable roof collected rainwater in Auckland, New Zealand. *Water Research* 35(6): 1518–1524.

Soller, J.A., Bartrand, T., Ashbolt, N.J., Ravenscroft, J.and Wade, T.J. (2010a). Estimating the primary etiologic agents in recreational freshwaters impacted by human sources of faecal contamination. *Water Research* 44(16): 4736–4747.

Soller, J.A., Eisenberg, J., DeGeorge, J., Cooper, R., Tchobanoglous, G. and Olivieri, A. (2006). A public health evaluation of recreational water impairment. *Journal of Water and Health* 4(1): 1–19.

Soller, J.A., Schoen, M.E., Bartrand, T., Ravenscroft, J. and Ashbolt, N.J. (2010b). Estimated human health risks from exposure to recreational waters impacted by human and non-human sources of faecal contamination. *Water Research* 44(16): 4674–4691.

SON (2007). *Nigerian Standard for Drinking Water Quality. Standard Organisation of Nigeria* http://www.unicef.org/nigeria/ng_publications_Nigerian_Standard_for_Drinking_Water_Quality.pdf (Accessed on October 2, 2014).

Sowunmi, K., Olamiji, O.M., Adesola, D.Y., Lawal, A.A., Adejoke, O.K., Emmanuel, K.A., Mohammadi, M.R. and Seun, O.O. (2022). *Antimicrobial Resistance of Escherichia coli Isolated from a Variety of Meats in Sabo Market, Ikorodu, Lagos, Nigeria.*

Spinks, J., Phillips, Robinson, S. and Van Buynder, P. (2006). Bushfires and tank rainwater quality. A cause for concern. *Journal of Water and Health* 4(1): 21–28.

Sprovieri, F. and Pirrone, N. (2008). Particle size distributions and elemental composition of atmospheric particulate matter in Southern Italy. *Journal of the Air and Waste Management Association* 58(6): 797–805.

StatsNz (2021) Natural sources of particulate matter. Natural sources of particulate matter I Stats NZ (Accessed on January 9, 2023).

Stolzenbach, K.D. (2006). Atmospheric deposition. UCLA Institute of the Environment and Sustainability. *Southern California Environmental Report Card, 2006.*

http://www.environment.ucla.edu/media/files/Atmospheric-Deposition-2006.pdf (Accessed on November 10, 2013).

Strachan, N.J.C., Doyle, M.P., Kasuga, F., Rotariu, O. and Ogden, I.D. (2005). Dose response modelling of *Escherichia coli* O157 incorporating data from foodborne and environmental outbreaks. *International Journal of Food Microbiology* 103(1): 35–47.

Sunnu, A., Afeti, G. and Resch, F. (2013). A long-term experimental study of the Saharan dust presence in West Africa. *Atmospheric Research* 87(1): 13–26.

Sunnu, A. K. (2006). An experimental study of the Saharan dust physical characteristics and fluxes near the Gulf of Guinea. Published PhD Thesis. Université du Sud Toulon-Var, France.

Swap, R., Garstang, M., Greco, S., Talbot, R. and Kallberg, P. (1992). Sahara dust in Amazon Basin. *Tellus* 44(B): 133–149.

Tegen, I., Harrison, S.P., Kohfeld, K., Prentice, I.C., Heinold, B., Helmert, J., Todd, M.C., Washington, R. and Dubovik, O. (2006). Modeling soil dust aerosol in the Bodèlè depression during the BODEX campaign. *Atmospheric Chemistry and Physics* 6(12): 4345–4359.

Tegen, I., Werner, M., Harrison, S.P. and Kohfeld, K.E. (2004). Relative importance of climate and land use in determining present and future global soil dust emission. *Geophysical Research Letters* 31(5): L05105.

Texas Water Development Board (2005). *The Texas Manual on Rainwater Harvesting*, 3rd Edition, Texas Water Development Board, Austin, TX. www.twdb.state.tx.us.

The James Hutton Institute (2019). Septic tanks as point source of pollutants and their impact on water quality. The Impact of Septic Tanks on Water Quality I The James Hutton Institute (Accessed on February 23, 2023).

Thomas, P.R. and Greene, G.R. (1993). Rainwater quality from different roof catchments. *Water Science and Technology* 28(3–5): 291–299.

Tiessen, H., Hauffe, H.K. and Mermut, A.R. (1991). Deposition of Harmattan dust and its influence on base saturation of soils in northern Ghana. *Geoderma* 49(3–4): 285–299.

Todd, M.C., Washington, R., Martins, J.V., Dubovik, O., Lizcano, G., M'Bainayel, S. and Engelstaedter, S. (2007). Mineral dust emission from the Bodélé depression, northern Chad, during BoDEx 2005. *Journal of Geophysical Research* 112(D6): D06207.

Tornevi, A., Axelsson, G. and Forsberg, B. (2013). Association between precipitation upstream of a drinking water utility and nurse advice calls relating to acute gastrointestinal illnesses. *PLoS ONE* [electronic resource] 8: 69918.

Uba, B.N. and Aghogho, O. (2000). Rainwater quality from different roof catchments in the Port Harcourt District, Rivers State, Nigeria. *Journal of Water Supply Research and Technology – Aqua* 49(5): 281–288.

Ukabiala, C. O., Nwinyi, Obinna, Abayomi, A. and Alo, B. I. (2010). Assessment of heavy metals in urban highway runoff from Ikorodu Expressway Lagos, Nigeria. *Journal of Environmental Chemistry and Ecotoxicology.* 2(3): 34–37. ISSN 2141 - 226X.

Umunnakwe, F. A., Idowu, E. T., Ajibaye, O., Etoketim, B., Akindele, S., Shokunbi, A. O., Otubanjo, O. A., Awandare, G. A., Amambu-Ngwa, A. and Oyebola, K. M. (2019). High cases of submicroscopic Plasmodium falciparum infections in a suburban population of Lagos, Nigeria. *Malaria Journal* 18: 433–441. doi: 10.1186/s12936-019-3073-7.

UN (2014). The United Nations world water development report 2014. The United Nations world water development report - Unesco Digital Library. United Nations (Accessed on February 21, 2023).

UN (2020). *Goal 6: Ensure Access to Water and Sanitation for All: Sustainable Development Goals. United Nations,* Water *and Sanitation* - United Nations *Sustainable* Development (Accessed on February 8, 2023).

UNEP (2012). *Sourcebook of alternative technologies for freshwater Augumentation in Latin America and the Caribbean.* United Nations Environment Programme. http://www.unep.or.jp/ietc/Publications/TechPublications/TechPub-8c/disinfect .asp?&session-id=5208e90404efcb82969f0cb87a73f49c (Accessed on November 15, 2015).

UNICEF and WHO (2021). *Protecting Child Rights in a Time of Crisis.* UNICEF annual report. UNICEF United Nations International Children's Emergency Fund (Accessed on February 21, 2023).

United Nations (2022). Take action today: The global goals. How to Achieve Sustainable Development Goals - The Global Goals (Accessed on May 30, 2023).

USAID (2012). Nigeria: Water and Sanitation profile. http://pdf.usaid.gov/pdf_docs/ PNADO937.pdf (Accessed on December 6, 2012).

Van Metre, P.C. and Mahler, B.J. (2003). The contribution of particles washed from rooftops to contaminant loading to urban streams. *Chemosphere* 52(10): 1727–1741.

Vasudevan, P., Tandon, M., Krishnan, C. and Thomas, T. (2001). Bacterial quality of water in DRWH. In: *Proceedings of the 10th International Rainwater Catchment Systems Conference.* International Rainwater Catchment Systems Association, Weikersheim, Germany, pp. 153–155.

Verstraeten, I.M., Fetterman, G.S., Sebree, S.K., Meyer, M.T. and Bullen, T.D. (2004). *Is Septic Waste Affecting Drinking Water from Shallow Domestic Wells Along the Platte River in Eastern Nebraska? USGS, Science for a Changing World.* fs07203l.pdf (usgs.gov) (Accessed on February 21, 2023).

Vialle, C., Sablayrolles, C., Lovera, M., Huau, M.C., Jacob, S. and Montrejaud-Vignoles, M. (2012). Water quality monitoring and hydraulic evaluation of a household roof runoff harvesting system in France. *Water Resources Management* 26(8): 2233–2241.

Vithanage, M., Bandara, P.C., Novo, L.A.B., Kumar, A., Ambade, B., Naveendrakumar, G., Ranagalage, M. and Magana-Arachchi, D.N. (2022). Deposition of trace metals associated with atmospheric particulate matter: Environmental fate and health risk assessment. *Chemosphere* 303(3): 135051. doi: 10.1016/j.chemosphere.2022.135051. PMID: 35671821.

Ward, S., Memon, F.A. and Butler, D. (2010). Harvested rainwater quality: The importance of appropriate design. *Water Science and Technology* 61(7): 1707–1714.

WaterAid (2015). Everyone, everywhere: A vision for water, sanitation and hygiene post-2015. http://www.wateraid.org/what-we-do/our-approach/research-and -publications/view-publication?id=b5203641-de28-4c9e-acf1-a511562ead45 (Accessed on November 25, 2015).

Weaver, L.J., Sousa, M.L., Wang, G., Baidoo, E., Petzold, C.J. and Kealing, J.D. (2015). A kinetic-based approach to understanding heterologous mevalonate pathway function in E. coli. *Biotechnology and Bioengineering* 112(1):111–119.

Weinmeyer, R., Norling, A., Kawarski, M. and Higgins, E. (2017). The Safe Drinking Water Act of 1974 and its role in providing access to safe drinking water in the United States. *AMA Journal of Ethics* 19(10): 1018–1026. doi: 10.1001/journal ofethics.2017.19.10.hlaw1-1710. PMID: 29028470.

Wellaware (2023). The Water Crisis | Well Aware (wellawareworld.org) (Accessed on February 18, 2023).

Whelan, G., Kim, K., Pelton, M.A., Soller, J.A., Karl, J., Castleton, K.J., Molina, M., Pachepsky, Y. and Zepp, R. (2014). An integrated environmental modelling framework for performing Quantitative Microbial Risk Assessments. *Environmental Modelling and Software* 55: 77–91.

WHO (2004). Water, sanitation and hygiene links to health. Available from: http://www.who.int/water_sanitation_health/facts2004/en/ (Accessed on October 7, 2015).

WHO (2011). *Guidelines for Drinking-Water Quality*. World Health Organisation, Geneva.

WHO (2016). Quantitative microbial risk assessment: Application for water safety management. WHO Document Production Services, Geneva, Switzerland. ISBN 978 92 4 1565370.

WHO/UNICEF (2019). Progress on household drinking water, sanitation and hygiene 2000–2017. World Health Organization (WHO) and United Nations Children's Fund (UNICEF). Available from: http://www.who.int/water_sanitation_health/publications/jmp-report2019/en/ (Assessed May 26, 2020).

WHO/UNICEF JMP (2012). WHO/UNICEF Joint Monitoring Programme for Water Supply and Sanitation estimates for the use of Improved Sanitation Facilities and Improved drinking water sources in Nigeria. http://wssinfo.org (Accessed on October 10, 2012).

WHO/UNICEF JMP (2014). *Progress on Drinking Water and Sanitation*, 2014 Update and MDG Assessment. World Health Organisation/United Nations Children Fund Joint Monitoring Programme for Water Supply and Sanitation. http://www.wssinfo.org/fileadmin/user_upload/resources/JMP_report_2014_webEng (Accessed on May 23, 2014).

WHO/UNICEF JMP (2015). *Progress on Sanitation and Drinking Water*. 2015 Update and MDG Assessment. World Health Organisation/United Nations Children Fund Joint Monitoring Programme for Water Supply and Sanitation. http://www.who.int/water_sanitation_health/publications/jmp_2015_update_compressed.pdf (Accessed September 28, 2015).

WHO and UNICEF (2019). For every child, reimagine. UNICEF annual report. UNICEF annual report. UNICEF United Nations International Children's Emergency Fund (Accessed on February 21, 2023).

World Bank (2013). Working for a world free of poverty. http:data.worldbank.org/country/Nigeria (Accessed on November 20, 2022).

WSMP (2008). Water and sanitation summary sheet *NIGERIA- country summary sheet. Water and Sanitation Monitoring Platform.* (Accessed on October 2, 2014). http://www.unicef.org/nigeria/ng_media_Water_sanitation_summary_sheet.pdf.

www.laportecountybeaches.com (Accessed on April 11, 2015).

www.moderndevice.com. *Product Code MD0550* (Accessed on September 20, 2014).

www.omega.co.uk/pptst/OM-CP-WIND101A-KIT.html (Accessed on November 28, 2014).

www.qmrawiki.canr.msu.edu (Accessed August 18, 2013).

Yang, K., LeJeune, J., Alsdorf, D., Lu, B., Shum, C.K. and Liang, S. (2012). Global distribution of outbreaks of water-associated infectious diseases. *PLoS Neglected Tropical Diseases* [electronic resource] 6: 1483–1495.

Yang, Y., Wang, H., Cao, Y., Gui, H., Liu, J., Lu, L., Cao, H., Yu, T. and You, H. (2015). The design of rapid turbidity measurement system based on single photon detection techniques. *Optics and Laser Technology* 73: 44–49.

Yaziz, M.I., Gunting, H., Sapari, N. and Ghazali, A.W. (1989). Variations in rainwater quality from roof catchments. *Water Research* 23(6): 761–765.

Zheng, X.G., Pu, J.H., Chen, R., Liu, X. and Shao, S. (2016). A novel explicit-implicit coupled solution method of SWE for long-term river meandering process induced by dambreak. *Journal of Applied and Fluid Mechanics* 9(6): 2647–2660.

Zhou, Z., Yang, Y., Li, X., Gao, W., Liang, H. and Li, G. (2012). Coagulation efficiency and flocs characteristics of recycling sludge during treatment of low temperature and micropolluted water. *Journal of Environmental Sciences* 24(6): 1014–1020.

Zhu, K., Zhang, L., Hart, W., Liu, M. and Chen, H. (2004). Quality issues in harvested rainwater in arid and semi-arid loss Plateau of Northern China. *Journal of Arid Environments* 57(4): 487–505.

Zmirou, D., Pena, L., Ledrans, M. and Letertre, A. (2003). Risks associated with the microbiological quality of bodies of fresh and marine water used for recreational purposes: Summary estimates based on published epidemiological studies. *Archives of Environmental Health* 58(11): 703–711.

Index